Physical Non-Linearities in Structural Analysis

International Union of Theoretical and Applied Mechanics

Physical Non-Linearities in Structural Analysis

Symposium Senlis, France
May 27–30, 1980

Editors
J. Hult and **J. Lemaitre**

With 109 Figures

Springer-Verlag
Berlin Heidelberg New York 1981

Prof. JAN HULT

Chalmers University of Technology
S-402 20 Gothenburg, Sweden

Prof. JEAN LEMAITRE

Laboratoire de Mécanique et Technologie
Université Paris VI · ENSET
61, Avenue du Président-Wilson
94230 CACHAN (FRANCE)

ISBN 3-540-10544-1 Springer-Verlag Berlin Heidelberg New York
ISBN 0-387-10544-1 Springer-Verlag New York Heidelberg Berlin

Offsetprinting: fotokop wilhelm weihert KG, Darmstadt · Bookbinding: Konrad Triltsch, Würzburg

2061/3020/543210

Preface

The Symposium was initiated in 1977 by Dr. M.G. CHABERT, Directeur des plans et programmes du C.E.T.I.M. His death, which occurred a few weeks before the Symposium, was deeply felt by all participants, who decided to dedicate this volume to his memory.

The purpose, as set out by Dr. CHABERT, was to bring together people speaking the language of research and the language of industry in discussing problems of structural analysis. The Centre Technique des Industries Mécaniques (C.E.T.I.M.) at Senlis was considered a suitable place for organizing such a meeting.

Modern structural analysis, aided by high speed computers, is mainly concerned with non-linear phenomena, such as plasticity, visco-plasticity, damage and fracture mechanics. Scientists involved in research in this field also speak two different languages: that of material analysts and that of structural analysts.

The Bureau of IUTAM had appointed the following Scientific Committee to implement the ideas proposed by Dr. CHABERT:

J. LEMAITRE, Université Paris VI, ENSET, Cachan, France Chairman

J. HULT, Chalmers University of Technology, Gothenburg, Sweden, Co-Chairman

G. CHABERT, Centre Technique des Industries Mécaniques, Senlis, France

F.A. LECKIE, University of Illinois, Urbana, Illinois, USA

G. MAIER, Istituto di Scienza e Technica delle Costruzioni, Polotechnico di Milano, Italy

Z. MROZ, Polish Academy of Sciences, Warsaw, Poland

G.S. PISARENKO, Academy of Sciences of UkrSSR, Kiev, USSR

The programme was arranged with plenary lecture sessions each morning on the following topics:

Constitutive Equations (Chairmen: M. SIMON and J. LEMAITRE)

Damage and Rupture Concepts (Chairman: Z. MROZ)

Numerical Formulation (Chairman: A. SAWCZUK)

Structural Analysis (Chairman: M. MORANVAL)

The afternoon were devoted to working sessions with short lectures aimed to support the discussion on the following subjects:

Strain, Damage and Rupture Behaviour (Chairman: J. HULT)

Numerical Methods and Procedures (Chairman: P. LADEVEZE)

Applications to Structures (Chairman: F.A. LECKIE).

The Scientific Committee wishes to thank all those who have contributed to make this Symposium a success:

C.E.T.I.M. through its Director, M. WIART, who was in charge of the organization,

the sponsors of the Symposium:

Société Francaise des Mécaniciens

Association Universitaire de Mécanique

Centre National de la Recherche Scientifique et Technique

Direction des Recherches et Etudes Techniques

Electricité de France,

the members of the Organizing Committee:

M. DUBOIS, M. CARROI, A. CHAUDOUET, J.M. BOISSENOT, J.L. CHABOCHE, R. DUBOIS, F. JUNG, P. LADEVEZE, M.P. LEMAITRE, G. LOUBIGNAC and D. SERRES

and finally all participants.

J. HULT   J. LEMAITRE

Participants

W. Alwis (Australia)
J.H. Argyris (FRG)
R. Billardon (France)
Z. Bilek (Czechoslovakia)
F. Billon (France)
S.R. Bodner (Israel)
J.M. Boissenot (France)
J.T. Boyle (UK)
M. Brunet (France)
G. Cailletaud (France)
M. Carroi (France)
J. Casier (France)
J.L. Chaboche (France)
A. Chadouet (France)
J.L. Chenot (France)
G. Coffignal (France)
A. Combescure (France)
G. Cordier (France)
L. Cristescu (France)
J.I. Curiskis (Australia)
Y.D. Dafalias (USA)
K. Dang Van (France)
O. Desbordes (France)
H. Di Benedetto (France)
S. Dordosz (Poland)
R.N. Dubey (Canada)
M. Dubois (France)
D. Eveno (France)
G.D. Galletly (UK)
A. Gajewski (Poland)
P. Grundy (Australia)
P. Hacq (France)
B. Halphen (France)
M. Hamada (France)
A. Hoffman (France)
J. Hult (Sweden)
I. Inoue (Japan)
S. Kaliszky (Hungary)
E. Krempl (USA)
P. Ladeveze (France)
J. Lambermont (Belgium)
F.A. Leckie (USA)
M.P. Lemaître (France)
J. Lemaître (France)
L. Librescu (Israel)
G. Loubignac (France)
J. Lubliner (USA)
D. Marquis (France)
J.B. Martin (South Africa)
M. Masur (USA)
M. Mazet (France)
R.M. McMeeking (USA)
D.G. Moffatt (UK)
M. Moranval (France)
D.N. Moreton (UK)
Z. Mroz (Poland)
S. Nagaki (Japan)
M. Nash (USA)
B. Nayroles (France)
S.Q. Nguyen (France)
E.T. Onat (USA)
P.D. Panagiotopoulos (Greece)
G. Pape (FRG)
Y. Partom (Israel)
J. Perrot (France)
P. Perzyna (Poland)
A. Piatti (Italy)
A.R.S. Ponter (UK)

M. Predeleanu (France)
A. Ricard (France)
J.C. Robinet (France)
M. Sandhu (FRG)
A.R. Santhakumar (India)
A. Sawczuk (Poland)
D. Serres (France)
F. Sidoroff (France)
M. Simon (France)
J. Suhara (Japan)
S. Tabata (Japan)
M. Tanaka (Japan)
J.M. Terriez (France)
G. Thierauf (FRG)
A. Turbat (France)
S. Valliapan (Australia)
Z. Waszczszyn (Poland)
L.E. Waz (FRG)
K.J. Willam (FRG)
Y. Yamada (Japan)
T. Yokobori (Japan)
J. Zarka (France)

Contents

Préambule inaugural

Maurice ROY
Membre de l'Institut.

Hôtes et Amis étrangers,

Chers Collègues,

Mesdames, Messieurs,

Je sais gré aux organisateurs du présent Symposium, donc principalement aux dirigeants du CETIM, de m'avoir invité à prononcer, en ce préambule inaugural, une sorte d'introduction générale aux travaux qui vont commencer.

Mais c'est le coeur serré par une profonde émotion que je dois, parce qu'il nous a - hélas - quittés au soir du premier jour de ce mois, évoquer d'abord la belle figure de l'instigateur de ce Symposium, Gilles CHABERT.

Esprit lumineux, Savant authentique non moins qu'authentique Homme de l'Art, se perfectionnant sans cesse en théorie comme en pratique, tel était Gilles CHABERT. On peut dire que sa disparition si prématurée a mis en deuil les sciences et industries mécaniques de FRANCE, qu'il servait de tout son immense talent au sein du CETIM depuis la fondation de cet organisme.

Ses qualités humaines, notamment la bienveillance et la générosité de son esprit, sa discrétion et sa modestie mais aussi sa fermeté d'âme, son courage et sa rigoureuse honnêteté conféraient à sa personnalité une exceptionnelle autorité morale, à la hauteur de sa non moins exceptionnelle valeur intellectuelle.

Ces qualités font comprendre la sincérité du poignant chagrin de tous ceux qui l'ont connu et, par suite, estimé et aimé.

S'il m'est permis d'évoquer un très personnel souvenir je révèlerai que je fis sa connaissance dans les ateliers d'un constructeur réputé de l'industrie automobile alors que, sorti brillamment de l'Ecole Normale Supérieure, ce Mécanicien de vocation avait choisi de venir apprendre en usine les conditions de la réalisation industrielle en Mécanique.

Cette amorce de carrière, probablement unique en son genre pour un Normalien scientifique, le prédestinait à faire bénéficier le CETIM, dès la naissance de celui-ci, de son savoir personnel très étendu et triple car associant science, technique et technologie.

Lorsqu'il me fit part de son projet initial du présent Symposium je me permis, tant pour en souligner la signification sur le plan international que pour faire honneur au CETIM, de lui suggérer que le CETIM veuille bien en demander l'inscription parmi les Symposiums officiellement patronnés par l'Union Internationale de Mécanique Théorique et Appliquée (en bref IUTAM) au Bureau de laquelle j'avais siégé pendant seize ans, de 1956 à 1972. A cette idée le CETIM souscrivit sans délai et j'ai pu appuyer efficacement cette requête, en particulier à l'Assemblée tenue par l'Union en 1978, laquelle y fit droit par un vote quasi-unanime.

Ainsi, le présent Symposium IUTAM-CETIM a normalement bénéficié d'un Comité Scientifique International dont, avec sa discrétion coutumière, Gilles CHABERT déclina la présidence, Comité que je salue ici et où je retrouve des Collègues très estimés parmi lesquels je salue particulièrement le très distingué Secrétaire Général de l'Union, Monsieur le Professeur Jan HULT, et le Président de ce Comité, Monsieur le Professeur Jean LEMAITRE.

Ce n'est évidemment pas ici le lieu de commenter davantage la carrière ni de détailler les titres et travaux du Membre, hélas disparu, de ce Comité dont j'ai tenu à évoquer d'abord la belle figure. Je sais d'ailleurs, que le CETIM tient à honneur de lui consacrer une Notice qui fera saisir en détail la diversité, l'étendue et la profondeur de ses mérites et je ne doute pas que la lumineuse et exemplaire mémoire de Gilles CHABERT inspire un fructueux déroulement du présent Symposium.

L'objet même de cette réunion a été nettement précisé dans la troisième circulaire du CETIM, à laquelle je me reporte maintenant.

Cet objet est la prise en compte, dans les calculs des structures, des NON-LINEARITES PHYSIQUES et, notamment des plus récents enseignements à leur égard des recherches qui les concernent. Ceci laisse expressément de côté les non-linéarités mathématiques ou géométriques, comme celles entraînées par des déformations cumulées et ordinairement appelées "grandes déformations".

Le sujet en question apparaît, par suite, d'un INTERET AUJOURD'HUI FONDAMENTAL POUR LE MECANICIEN, car toute structure doit être calculée en faisant état des meilleures connaissances acquises sur les PROPRIETES MECANIQUES REELLES DES MATERIAUX considérés. La difficulté essentielle reste aujourd'hui la traduction, à l'ECHELLE DU CONTINU MACROSCOPIQUE desdites propriétés sur lesquelles les connaissances les plus fines sont, ou seront désormais, principalement acquises par des recherches opérées à l'ECHELLE DITE MICROSCOPIQUE et le plus souvent relative à la MATIERE DISCONTINUE.

On sait que les principes des méthodes d'approximation mathématique applicable à la résolution des équations de la Mécanique des solides (et donc des structures) ont été établis dès la fin du siècle dernier ou le début du présent.

Mais, en dehors de cas très simples et trop particuliers, l'utilisation de telles méthodes de calcul restait alors subordonnée à l'avènement de machines à calculer capables d'exécuter très rapidement des programmes de calcul très complexes.

Depuis trois décennies environ et faisant suite à la découverte des propriétés des corps semi-conducteurs et aux progrès subséquents de composants électroniques très divers et de plus en plus miniaturisés, les méthodes de calcul par éléments finis ou par différences finies, voire des méthodes combinées, ont suscité une prolifération de travaux foncièrement mathématiques, où la formation de LANGAGES et de PROGRAMMES appropriés à ces extraordinaires machines a pris largement le pas sur la caractérisation intrinsèque des matériaux réels, à laquelle le présent Symposium contribuera, nous l'espérons tous, à restituer son importance essentielle.

Et c'est pourquoi j'en souhaite, pour ma part, un plein succès.

Les NON-LINEARITES PHYSIQUES en question ont aussi une interprétation énergétique dont je tiensà souligner l'importance. Cette INTERPRETATION ENERGETIQUE doit s'entendre des transformations et dissipations d'énergie qui accompagnent la mise en jeu desdites non-linéarités. Depuis plus d'un demi-siècle j'ai toujous attiré l'attention des ingénieurs et techniciens de la Mécanique - théoriciens aussi bien qu'expérimentateurs - sur l'importance, aux fins de compréhension et d'explication, d'une telle "interprétation énergétique", intimement liée à la DEFINITION DE L'ETAT INTERIEUR de tout élément matériel du continu macroscopique en cause. La dissipation d'énergie, dès qu'elle existe, exerce inévitablement un effet cumulatif sur ledit état, ce qui ne peut manquer de retentir à la longue sur le comportement ultérieur du matériau considéré.

Cette voie interprétative des phénomènes réels est aussi la plus naturelle pour tenir compte aisément, s'il y a lieu, de conditions de flux thermique et/ou de température à la frontière des systèmes en cause, ce qui concerne beaucoup de structures modernes comprenant notamment les machines thermiques et les appareils et machines thermochimiques.

Enfin, l'interprétation énergétique fournit une voie d'accès à la compréhension et à l'explication des transformations chimiques ou physicochimiques des matériaux de plus en plus divers qui constituent les structures à calculer.

Laissant cette digression et revenant au programme de vos travaux, il semble très probable que les "sessions de travail" des trois après-midi de ce jour, de demain et d'après-demain seront très chargées en échange de vues et discussions sur des points très divers, voire singuliers, du programme dont le cadre général a été judicieusement et nettement délimité. Ce pourra donc être une lourde charge pour les présidents de ces "Working Sessions" que de diriger ou d'organiser sans contraintes indésirables ces discussions et d'en circonscrire le cadre pour éviter une dispersion trop large, nuisible alors à la pertinence et à l'efficacité.

A cet égard et tout en souhaitant patience et courage aux présidents investis de cette charge, je souhaite aussi clarté et concision aux très nombreux intervenants.

C'est ainsi que pourront être recueillis, pour les faire mûrir ensuite, les meilleurs fruits du travail d'amicale coopération qui va s'engager.

Numerical Techniques for the Finite Element Analysis of Elastic and Inelastic Material Nonlinearities

J. H. ARGYRIS, L. E. VAZ and K. J. WILLAM

Institut für Statik und Dynamik der Luft- und Raumfahrtkonstruktionen
University of Stuttgart, Pfaffenwaldring 27, D-7000 Stuttgart 80, W. Germany

Summary

The objective of the paper is to unify nonlinear elastic and inelastic computations within the framework of the finite element method. Essentially two aspects are considered, (i) the step-by-step integration of the inelastic rate equations, and (ii) the solution of the underlying algebraic problem within each time step. In the first area stability and accuracy are of prime concern for the integration of the inelastic evolution laws with explicit or implicit time operators. In the second field direct forward approaches are contrasted with iterative strategies using both initial load and tangential gradient methods for advancing the solution within a given time interval. The overall computational schemes are illustrated at the viscoplastic example of a thick-walled cylinder subjected to internal pressure.

1. Introduction

The finite element analysis of inelastic structures involves the numerical integration of hereditary or evolutionary type constitutive equations. From the computational viewpoint the rate formulation of inelastic evolution laws offers the decisive advantage that memory effects are included automatically in the initial conditions of the underlying incremental strategy rather than in convolution integrals which require storage and consideration of the entire history. Therefore, the recent developments of inelastic constitutive equations concentrates on the identification of appropriate growth laws either directly for strain or internal variables which represent the microstructural changes in the material. In most cases, the new constitutive rate equations do no longer decouple inelastic mechanisms e.g. into creep and plasticity, they rather resort to a coupled formulation in terms of inelastic strain or even total strain.

2. Time operators

In a previous paper [1] the authors studied different numerical strategies for the solution of inelastic structural problems. Two families of linear $p=1$ and cubic $p=3$ Hermitean operators were presented and applied to the time approximation of inelastic evolution laws of viscous flow, viscoelastic and viscoplastic material behaviour whereby the higher order operator increased substantially the overall accuracy at little additional expense. Further control was introduced by the sampling parameter $0 \leq \xi \leq 1$ which designates the collocation point within the time step Δt and encompasses the explicit forward $\xi = 0$ scheme up to the fully implicit backward $\xi = 1$ method.

The algorithms were examined with regard to accuracy and stability as well as the convergence behaviour of the underlying iterative solution when an implicit formulation is used. In summary, the midstep rules with $\xi = 1/2$ yield a higher order consistency than the alternative $\xi \neq 1/2$ sampling strategies, whereby the linear midstep scheme with $p=1$, $\xi = 1/2$ exhibits oscillatory tendencies in contrast to the cubic midstep operator. It is widely known that the stability properties improve in general with increasing ξ-values (for example for $\xi \geq 1/2$ unconditional stability is attained in the case of linear viscoelastic evolution laws). Clearly, the overstable backward operator with $\xi = 1$ yields the most insensitive predictions, although at the cost of high numerical damping and thus degrading accuracy.

3. Computational schemes

It was rather intriguing that specific material properties such as aging and strain-hardening improved the stability behaviour of the explicit forward operator. Similar effects could be observed if the convergence properties of e.g. the predictor-corrector strategy were studied for the iterative initial load solution of implicit $\xi > 0$ algorithms. The convergence behaviour however deteriorates with increasing ξ-values and leads in the worst case of a simple viscoplastic overstress model to time step restrictions $\Delta t < \Delta t_c$ which are of the same order as the stability limit Δt_s of the corresponding explicit operator. Therefore, the computation of the long term response regime or steady-state conditions requires the tracing of the entire transient regime with a highly refined time step strategy, irrespectively if a simple forward scheme is used or a predictor-corrector method for the iterative solution of an implicit algorithm.

In order to overcome these inherent convergence difficulties the Newton-Raphson method was introduced in [2] which enlarges the range and rate of convergence of the iterative solution method at the cost of repeated updating and factorizing a gradient stiffness matrix.

For this reason the Quasi-Newton methods are adopted below for the iterative implementation of the implicit time operators resorting to rank two modifications of the inverse stiffness within the framework of the traditional initial load method. In this way the costly formation and factorization of structural gradient matrices is avoided altogether. As a matter of fact, the initial elastic stiffness is retained in the proposed scheme during the entire inelastic solution process, whereby the inelastic updates are converted into equivalent pseudo-loads. In this context, the BFGS-algorithm of Broyden-Fletcher-Goldfarb-Shanno [3] is successfully applied to the iterative solution of the implicit time marching scheme. The convergence problems of the predictor-corrector strategy are essentially resolved by the BFGS algorithm such that "arbitrarily large" time steps can be utilized e.g. for the asymptotic steady state elastoplastic solution via "equivalent" viscoplastic analysis (at least in conjunction with the fully implicit $\xi = 1$ operator).

The iteration procedure may be bypassed altogether by the forward gradient scheme [4, 5] in which a truncated Taylor-series expansion around the initial conditions is used to predict the response behaviour within the time interval Δt . This leads to a direct forward strategy along the line of the traditional explicit initial load method except for the additional gradient correction of first order. This "tangential" approach is unconditionally stable, however the accuracy is rather limited because of the "local" character of the Taylor expansion. The numerical damping of the basic time stepping strategy with $\xi = 1$ may be improved by the $\xi = 1/2$ weighting scheme, although at the expense of oscillatory tendencies. Moreover, the recent proposals for more sophisticated evolution laws of inelastic behaviour [6] lead to non-symmetric tangential stiffness matrices which unduly complicate the computational procedure.

For illustration different aspects of the BFGS-approach are studied below for the iterative solution of the implicit midstep and backward operators at the example of a pressurized cylinder. The computational implications are compared with those of the traditional initial load methods with and without iterative correction. For completion the forward gradient technique is applied and compared with the iterative BFGS solution of implicit time operators which does not require the formation and factorization of a tangential stiffness matrix.

4. Viscoplastic example

For illustration we consider the simple example of thick-walled cylinder subjected to internal pressure. Fig. 1 shows the basic lay-out and the response behaviour of the underlying viscoplastic overstress model. The short-term response with $t \to 0$ corresponds to the purely elastic Lamé solution, while the long-term response with $t \to \infty$ agrees with the steady state elasto-plastic solution of Prager and Hodge.

4.1 Initial load strategies

In the simplest case the inelastic material process is converted entirely into equivalent loads which constrain the elastic response path. This splitting of the structural operator

$$\mathbf{S}(\dot{\mathbf{r}}, \mathbf{r}) = \mathbf{R} \tag{1}$$

into an elastic stiffness and inelastic load contributions

$$\mathbf{S}(\dot{\mathbf{r}}, \mathbf{r}) = \mathbf{K}\,\mathbf{r} - \mathbf{J}(\dot{\mathbf{r}}) \tag{2}$$

leads to the following rate formulation of equilibrium between elastic restoring forces, external loads and inelastic dissipation forces

$$\mathbf{K}\,\dot{\mathbf{r}} = \dot{\mathbf{R}} + \dot{\mathbf{J}}_{\eta} \tag{3}$$

This finite element formulation of structural equilibrium results immediately from the additive decomposition of total strain rate $\dot{\boldsymbol{\gamma}}$ into elastic $\dot{\boldsymbol{\varepsilon}}$ and in inelastic

$\dot{\boldsymbol{\eta}}$ strain rate components whose evolution is controlled by the constitutive relations

$$\dot{\boldsymbol{\sigma}} = \mathbf{E}(\dot{\boldsymbol{\gamma}} - \dot{\boldsymbol{\eta}}) \qquad \text{where} \quad \boldsymbol{\sigma} = \mathbf{L}\boldsymbol{\sigma}_d \tag{4}$$

$$\dot{\boldsymbol{\eta}} = \mathbf{f}(\boldsymbol{\sigma}) = \gamma \langle \Phi \rangle \boldsymbol{\sigma}_d \qquad \text{and} \quad \Phi = \frac{3J_2}{\sigma_y^2} - 1$$

The differential formulation is now discretized in the time domain for example by a piecewise linear approximation of the inelastic strain within the normalized time interval $0 \leq \xi \leq 1$

$$\boldsymbol{\eta}_\xi = (1-\xi)\boldsymbol{\eta}_0 + \xi\boldsymbol{\eta}_1 \quad \text{and} \quad \dot{\boldsymbol{\eta}}_\xi = \frac{\boldsymbol{\eta}_1 - \boldsymbol{\eta}_0}{\Delta t} \tag{5}$$

which yields an expression for the inelastic strain increment as follows

$$\boldsymbol{\eta}_\Delta = \Delta t\, \mathbf{f}_\xi \qquad \text{where} \quad \mathbf{f}_\xi = \mathbf{f}(\boldsymbol{\sigma}_\xi) \tag{6}$$

It is intriguing that the higher order cubic Hermitean approximation introduced in [1, 7] leads to an analogous expression for the inelastic strain increment

$$\boldsymbol{\eta}_\Delta = \Delta t\,(c_0\mathbf{f}_0 + c_\xi\mathbf{f}_\xi + c_1\mathbf{f}_1) \tag{7}$$

which involves only two additional function evaluations $\mathbf{f}_0, \mathbf{f}_1$ at the beginning and end of the time increment Δt .

Clearly the explicit forward strategy with $\xi = 0$ for the incremental solution of eq. (3)

$$\mathbf{K}\,\mathbf{r}_\Delta = \mathbf{R}_\Delta + \mathbf{J}_\Delta \qquad \text{where} \quad \mathbf{J}_\Delta = \int_V \delta\boldsymbol{\gamma}^t\,\mathbf{E}\,\boldsymbol{\eta}_\Delta\, dV \tag{8}$$

is identical with the direct initial load method without iteration. The implicit $\xi > 0$ weighting schemes require in general iteration for example in form of a simple predictor-corrector scheme (successive substitution method). The underlying stability and convergence study furnishes time step restrictions for the linear expansion of viscoplastic strain which are shown in Fig. 2 where χ denotes the spectral radius of the material matrix $\mu\mathbf{L}\mathbf{E}$ and Φ the amount of overstress. Increasing ξ-values stabilize on one hand the time operator, on the other hand the convergence properties deteriorate. Therefore, the implicit time operators offer no advantage over the direct forward initial load approach if viscoplastic growth laws are used. Fig. 3 illustrates the history of equivalent stress at the inside wall. The prediction of the direct forward and the iterative midstep and backward algorithms exhibits little differences because of the small time steps which are necessary in order to satisfy the stability and convergence criteria which depend primarily on the amount of overstress.

4.2 Gradient strategies

The convergence difficulties of the predictor-corrector scheme are resolved if we resort to the more elaborate Newton-Raphson iteration scheme which requires repeated formation and factorization of structural gradient matrix $\mathbf{K}_T$ [2]. The underlying iteration method for a given time step $\Delta t = t_1 - t_0$

$$\mathbf{K}_T^i \, d\mathbf{r}^i = \mathbf{F}(\mathbf{r}^i) \tag{8}$$

localizes the roots of the residual load $\mathbf{F}(\mathbf{r}^i)$ due to constitutive incompatibilities, i.e. the error in the incremental material law $\dot{\boldsymbol{\eta}} = \mathbf{f}(\boldsymbol{\sigma})$

$$\mathbf{F}(\mathbf{r}^i) = \int_V \partial\boldsymbol{\gamma}^t \left[(\mathbf{E}_T^i)^{-1} (-\boldsymbol{\gamma}_\Delta^i + \Delta t \, \mathbf{f}_\xi^i + \mathbf{E}^{-1} \boldsymbol{\sigma}_\Delta^i) - \boldsymbol{\sigma}_1^i \right] dV + \mathbf{R}_0 \tag{9}$$

The tangential material law $\mathbf{E}_T$ is now composed of an elastic and an inelastic contribution involving the gradient of the inelastic evolution law

$$\mathbf{E}_T^i = \left(\mathbf{E}^{-1} + \Delta t \, \xi \, \frac{\partial \mathbf{f}_\xi^i}{\partial \boldsymbol{\sigma}} \right)^{-1} \quad \text{where} \quad \frac{\partial \mathbf{f}_\xi}{\partial \boldsymbol{\sigma}} = \frac{\partial}{\partial \boldsymbol{\sigma}} \left[\mathbf{f}(\boldsymbol{\sigma}_\xi) \right] \tag{10}$$

Clearly, symmetry is maintained only in special cases such as the viscoplastic overstress model with an associated flow rule. In contrast, novel constitutive proposals such as the total strain viscoplastic law in [6] lead to a non-symmetric tangential stiffness approach which complicates considerably the overall computations. For this reason the Quasi-Newton BFGS-method is applied below which generates a series of approximate gradient stiffnesses in the form of a variable metric search procedure

$$\mathbf{r}^{i+1} = \mathbf{r}^i + s^i \mathbf{d}^i \tag{11}$$

The BFGS-algorithm [3] maintains symmetry and updates the structural gradient matrix - or rather its inverse - by rank two modifications in the form of

$$(\bar{\mathbf{K}}_T^{i})^{-1} = (\mathbf{I} + \mathbf{w}_i \mathbf{v}_i^{t}) \, (\bar{\mathbf{K}}_T^{i-1})^{-1} (\mathbf{I} + \mathbf{v}_i \mathbf{w}_i^{t}) \qquad (12)$$

where the operations with the two intermediate vectors $\mathbf{v}_i$ and $\mathbf{w}_i$ only involve inner vector products and vector matrix products on the right-hand side. Computationally this is equivalent to the initial load iteration without changes of the structural stiffness matrix within the time step.

Fig. 4 shows the history of the equivalent stress at the inside wall for the implicit midstep and backward algorithms. The constant time step strategy with $\Delta t = 1$ clearly exceeds the critical values for stability and convergence of the traditional initial load method. It should be noted that the elastic stiffness was retained during the entire solution process whereby satisfactory convergence was attained within two cycles of iterations.

The distribution of equivalent stress across the wall is illustrated in Figs. 6 and 7 for the large time step of $\Delta t = 10$. It indicates clearly the oscillatory tendencies of the midstep algorithm while the backward operator exhibits sufficient numerical damping to approach monotonically the steady state elasto-plastic condition within a single time step. As a matter of fact, the steady state condition was attained with a single time step of $\Delta t = 1000$ and $\gamma = 1$.

In completion the forward gradient technique is also considered [4, 5] which corresponds to a modified Newton-Raphson algorithm without iteration. In this case the tangential stiffness matrix $\mathbf{K}_T$ is formed and factorized only once at the beginning of each time step with the compound elastic-viscoplastic material law

$$\mathbf{E}_T = \left(\mathbf{E}^{-1} + \Delta t \gamma \frac{\partial \mathbf{f}_o}{\partial \boldsymbol{\sigma}}\right)^{-1} \quad \text{where} \quad \frac{\partial \mathbf{f}_o}{\partial \boldsymbol{\sigma}} = \frac{\partial}{\partial \boldsymbol{\sigma}} [\mathbf{f}(\boldsymbol{\sigma}_o)] \qquad (13)$$

This computational strategy results from tangential linearization at a point of the solution path in the spirit of a truncated Taylor's expansion [5]. The gradient operator infers an implicit time operator inspite of the direct forward scheme similar to the Hermitean approximation of the solution path within a finite time interval. In the case of the forward gradient operator stability and convergence are of no concern, however accuracy is a problem because of the underlying linearization and the finite size of the time interval.

Fig. 5 shows again the history of equivalent stress at the inside wall for the forward gradient technique with $\xi=1/2$ and $\xi=1$. The solutions in the transient regime do not compare as well as those of the Hermitean operators in Fig. 4, however, the asymptotic behaviour converges again rapidly towards the steady state elasto-plastic solution. Figs. 6 and 7 show for the larger time step of $\Delta t=10$ that the forward gradient algorithms are performing rather poorly. As a matter of fact, for a time step of $\Delta t=1000$ large oscillations could be observed in the distribution of equivalent stress in the cylinder wall for both the $\xi=1/2$ and the $\xi=1$ strategy because of the discontinuous loading condition in viscoplasticity which is here controlled by the overstress at the beginning of the time step.

5. Conclusion

The BFGS-iteration method takes full advantage of the improved stability properties of implicit time operators without introducing convergence limitations in the viscoplastic analysis. Computationally this Quasi-Newton algorithm compares with the traditional initial load iteration method and overcomes the convergence problems of the simple predictor-corrector approach even if the elastic stiffness is retained as reference stiffness during the entire time marching process. In this case the inelastic updates of the tangential stiffness matrix are converted into equivalent loads in form of successive modifications of rank two.

Together with the fully implicit backward operator the BFGS-method is well suited for single step predictions of the steady state condition with very large time steps. In the case of proportional loading the viscoplastic long-term behaviour coincides with the corresponding elasto-plastic solution, however, path dependence precludes a more general statement to this end.

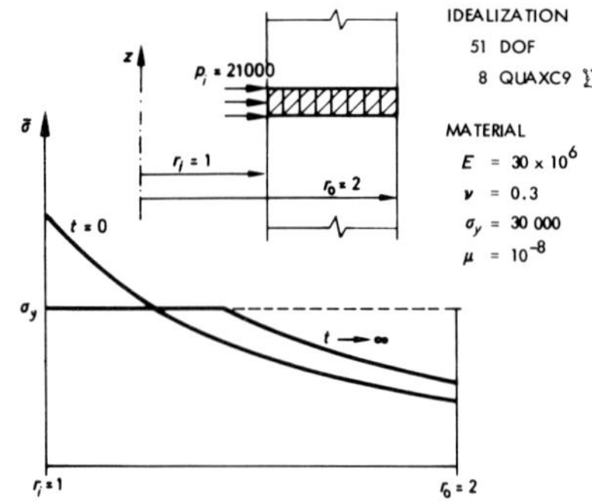

Fig. 1. THICK-WALLED CYLINDER
VISCOPLASTIC OVERSTRESS MODEL

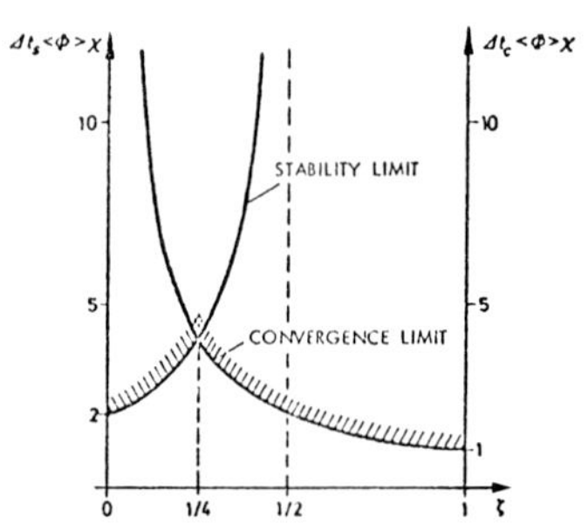

Fig. 2. TIMESTEP RESTRICTION FOR LINEAR OPERATOR

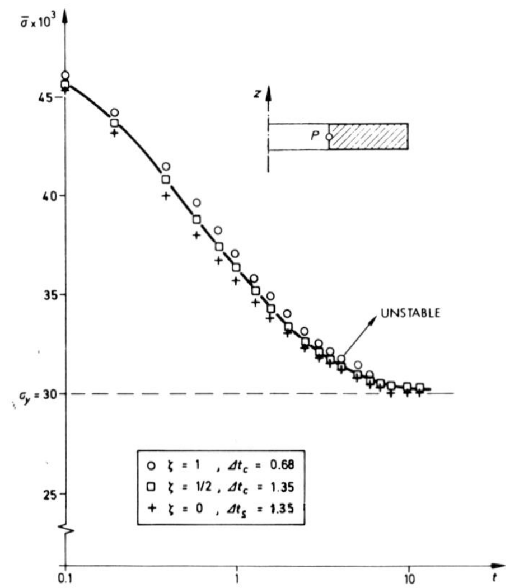

Fig. 3. THICK-WALLED CYLINDER, DIRECT AND ITERATIVE INITIAL LOAD METHODS
EQUIVALENT STRESS HISTORY AT POINT P

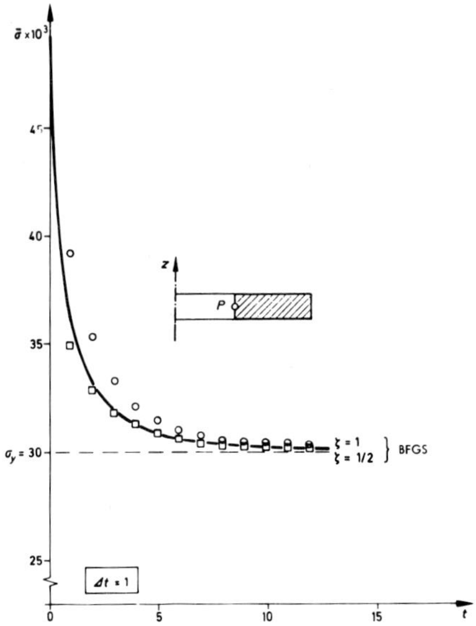

Fig. 4. THICK-WALLED CYLINDER, ITERATIVE BFGS-METHODS
EQUIVALENT STRESS HISTORY AT POINT P

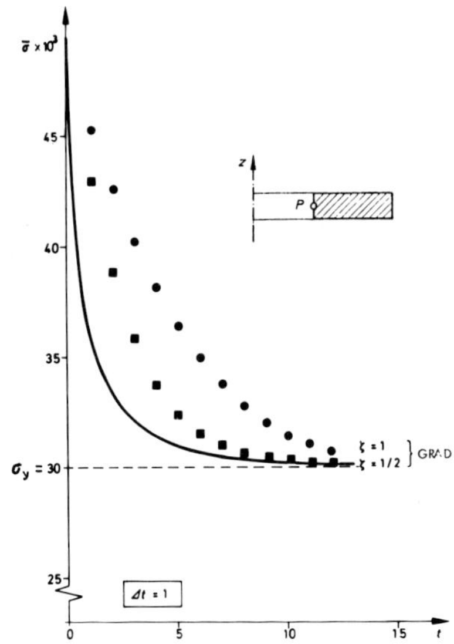

Fig. 5. THICK-WALLED CYLINDER, FORWARD GRADIENT METHODS
EQUIVALENT STRESS HISTORY AT POINT P

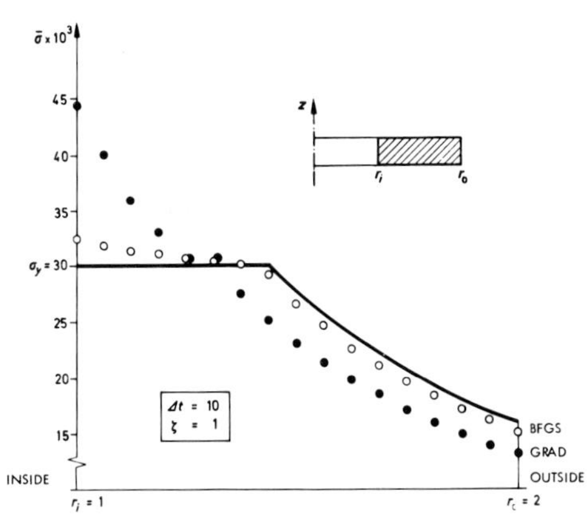

Fig. 6. THICK-WALLED CYLINDER, ITERATIVE BFGS AND FORWARD GRADIENT METHODS
EQUIVALENT STRESS DISTRIBUTION ACROSS CYLINDER WALL, $t = 10$

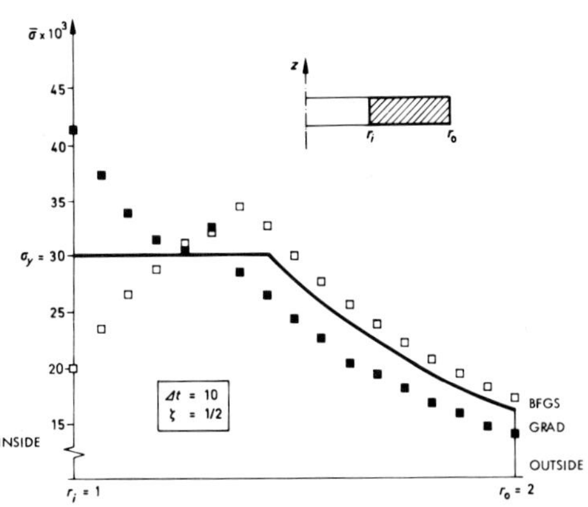

Fig. 7. THICK-WALLED CYLINDER, ITERATIVE BFGS AND FORWARD GRADIENT METHODS
EQUIVALENT STRESS DISTRIBUTION ACROSS CYLINDER WALL, $t = 10$

References

[1] J.H. Argyris, L.E. Vaz, K.J. Willam, Improved solution methods for inelastic rate problems, Comp. Meth. Appl. Mech. Eng. 16 (1978) 231-277.

[2] T.J.R. Hughes, R.L. Taylor, Unconditionally stable algorithms for quasi-static elasto-viscoplastic finite element analysis, Comp. Struct. 8 (1978) 169-173.

[3] H. Matthies, G. Strang, The solution of nonlinear finite element equations, Int. J. Num. Meth. Eng. 14 (1979) 1613-1626.

[4] K.J. Willam, Numerical solution of inelastic rate processes, Comp. Struct. 8 (1978) 511-531.

[5] M.B. Kranchi, O.C. Zienkiewicz, D.R.J. Owen, The visco-plastic approach to problems of plasticity and creep involving geometric non-linear effects, Int. J. Num. Meth. Eng. 12 (1978) 169-181.

[6] R.M. Zirin, E. Krempl, A finite element time integration method for the theory of viscoplasticity based on infinitesimal total strain, ASME Publ. 80-C2/PVP-33 (1980).

[7] J.H. Argyris, J.St. Doltsinis, W.C. Knudson, L.E. Vaz, K.J. Willam, Numerical solution of transient nonlinear problems, Fenomech '78 Proc., University of Stuttgart, Aug. 30 - Sept. 1, 1978, Comp. Meth. Appl. Mech. Eng. 17/18 (1979) 341-409.

Elastic-Plastic Fracture Mechanics of Cantilever Beam Specimen

Z. BÍLEK, S. PTÁK *, Z. KNÉSL

Institute of Physical Metallurgy, Czechoslovak Academy of Sciences, Žižkova 22, 616 62 Brno, Czechoslovakia

*National Research Institute for Machine Design
250 97 Prague-Běchovice, Czechoslovakia

Introduction

In recent years, the finite element method (fem) has been employed for analysis of structures with cracks, and many successful applications have been made (see rewiew papers [1-3]) for fracture mechanics specimens exhibiting non--linear material behaviour. The present paper is concerned with elastic-plastic finite element investigations of rectangular double cantilever beam (rdcb) specimen shown in Fig. 1a. This type of wedge-loaded test specimen is frequently used for crack initiation, propagation and arrest studies in brittle and ductile structural materials [4,5]. When stress intensities and dynamic toughnesses are predicted from measured displacements and crack lengths using the approximate one dimensional models [5-7] based on beam theory, there are two major sources of error: (1) the models assume the load acts perpendicular to the crack line. Wedge loading introduce a component parallel to the crack causing a perturbation of the displacements near the loading points (2) the models assumes that the applied load is distributed uniformly over the beam height which is not actually true because the load is applied via pins to holes in the specimen arms, - Fig. 1. By appropriate fem treatment we remove the both error sources and the validity of approximate approaches [5-7] to the rdcb specimen mechanics can be assesed. Our results are too extensive to allow discussion, in a paper of the length intended, so only the most interesting data are summarized and the reader is referred for further information to [8].

Analysis and Results

The dcb specimen was represented by 697 triangular constant strain elements and 395 nodes essentially focused around the crack tip. The smallest crack tip elements have dimensions of 0.01a where $a = 30$ mm is the crack length. The stress-strain Ramberg-Osgood relationship was assumed for all calculations with strain hardening exponent $n = 11$, yield strength at 2% set $\sigma_0 = 240$ MPa and Young's modulus $E = 2.1 \times 10^5$ MPa.

The stress intensity factor K_I was obtained by stress component σ_{yy} extrapolation to the crack tip. To test the validity of these results comparison with the stress intensity factors calculated according various dcb specimen beam models [6] was made in Table I. Fig. 1b shows the plastic zone devepment, for small and large scale yielding as well.

Table I

$K_I bh^{3/2} / \sqrt{12}$ Pa						
fem	bound.coll. [6]	Euler beam theory [6]		Timoshenko beam theory		
		Rigid found.	Winkler found.	Rigid found.	Gener. found.	Winkler found.
2.16	2.37	1	2.24	1.41	2.24	2.63

The J - integral [1,2] was computed from the stress, strain and displacement fields produced by fem along two semicircles surrounding the crack tip. The calculated dependence of J on dimenssionless load is plotted together with the set of independent experimental measurements performed at different temperatures in Fig. 2. The good agreement of both is evident. Fig. 3a, b demonstrate the dependence between the clip gage displacement 2Δ defined at 4.5 mm from the specimen surface,

the load points displacement 2δ and applied load. The experimental measurements are bounded by plane strain and plane stress solutions.

REFERENCES

[1] Rice, J.R.: In: The Mechanics of Fracture, AMD 19, 1977, 23.

[2] Rice, J.R. et al.: In: Proc. of FENOMECH Conference, Stuttgart, 1978, 21.

[3] Gallagher R.H.: In: Proc. of the first conference on numerical methods in fracture mechanics, Swansee, 1978, 3.

[4] Hoagland, R.G. et al.: In: ASTM STP 627, 1977, 177.

[5] Bílek, Z.: Scripta Metallurgica 12, 1978, 1101.

[6] Kanninen, M.F.: Int. J. Fracture, 10, 1974, 415.

[7] Kanninen, M.F. et al.: In: ASTM STP 627, 1977, 19.

[8] Bílek, Z. et al.: Eng. Fracture Mechanics, in preparation

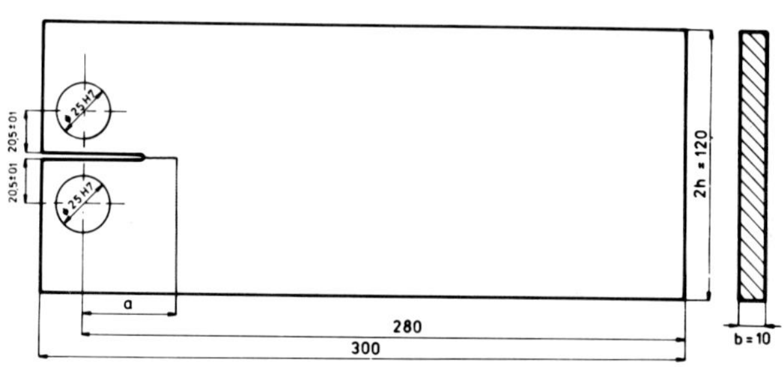

Fig. 1a: Geometry and dimensions of rdcb specimen analysed in this paper.

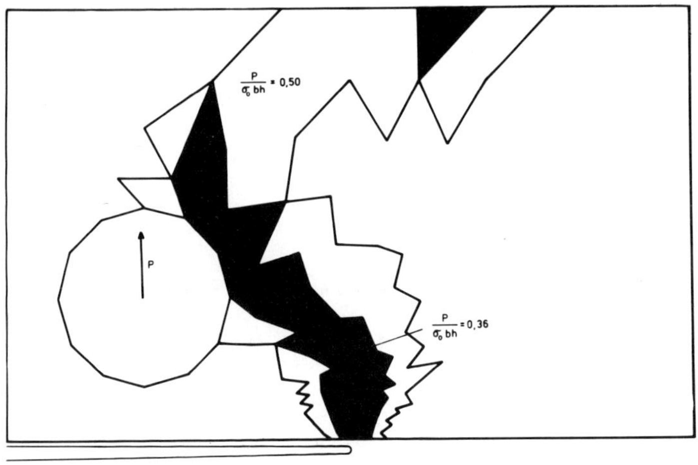

Fig. 1b: Development of the plastic zone for plane strain at two load levels is shown.

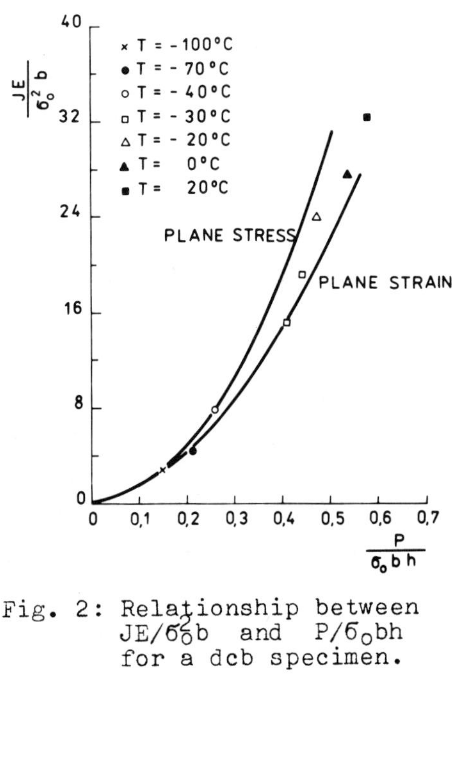

Fig. 2: Relationship between $JE/\sigma_0^2 b$ and $P/\sigma_0 bh$ for a dcb specimen.

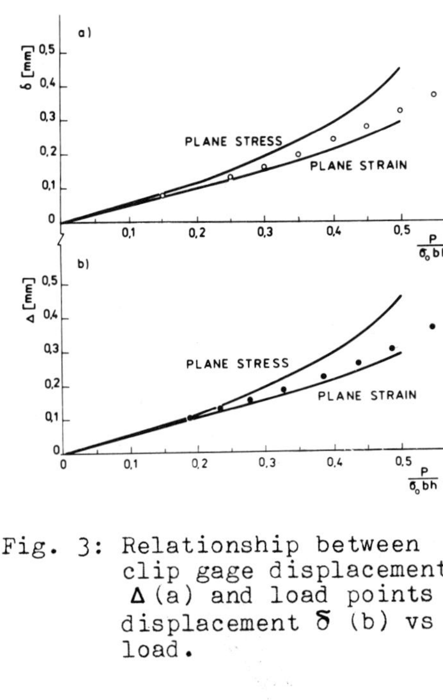

Fig. 3: Relationship between clip gage displacement Δ (a) and load points displacement δ (b) vs load.

A Procedure for Including Damage in Constitutive Equations for Elastic-Viscoplastic Work-Hardening Materials

S.R. BODNER

Faculty of Mechanical Engineering
Technion - Israel Institute of Technology, Haifa, Israel.

Abstract
The elastic-viscoplastic work-hardening constitutive equations of Bodner-Partom are modified to include damage as an internal state variable. An essential part of the formulation is the development of evolution equations for the damage parameter. This is done initially for the case of isotropic damage and isotropic work-hardening using methods similar to those proposed by Leckie and Hayhurst. A modification of the method is the assumption of a functional form for the damage parameter itself rather than for its time derivative. The formulation is extended to include nonisotropic damage and work-hardening using a procedure similar to that developed by Stouffer and Bodner for anisotropic work-hardening.

Introduction

This paper is concerned with the incorporation of damage as an internal state variable in a set of constitutive equations for elastic-viscoplastic, work-hardening materials. The damage parameter is intended to represent the degradation in the ability of the material to support stress (in the sense of Kachanov [1]). The other internal state variable in the equations is a parameter which represents resistance to plastic flow, i.e., work-hardening. An essential part of the formulation is the development of evolution equations for the damage parameter.

The reference set of constitutive equations are taken to be those of Bodner and Partom [2,3], which do not require a yield criterion or loading/unloading conditions. These were originally developed for isotropic hardening [3], and were subsequently extended to uniaxial cyclic loading in [4], and then for general loading histories in [5]. In the general case, [5], the material develops anisotropic hardening characteristics, i.e., induced plastic anisotropy, and also exhibits volume changes due to plastic deformation. Another development of the theory [6] involved the introduction of a hardness recovery term which was essential to predict secondary creep. Anelastic effects are absent in the formulations described in [2-6] but could, in principle, be added in a manner similar to that described in [7]. Anelastic effects in metals are primarily observed as transient effects at high temperatures. However, they are usually secondary factors at the lower temperatures and also for reasonably steady state conditions at high temperatures.

Inclusion of damage in the constitutive equations would be important if the equations are intended to represent material response characteristics leading to material failure, e.g. tertiary creep. Coupling the damage concept with the material response equations seems to follow the physical situation more closely than introducing independent and uncoupled failure criteria. In the context of the coupled equations, failure would be indicated by a breakdown in the material response behavior.

A particular procedure for introducing damage as an internal state variable in the equations of [3-6] is proposed in this paper. The damage parameter is assumed to have the properties of a special second order tensor similar to those of the general work-hardening variable described in [5]. A form for the associated evolution equation is also proposed.

Formulation of Equations

A. Isotropic Case

The inclusion of damage into the constitutive equations will first be described for the isotropic case, i.e., where damage is represented by a single scalar variable, and then generalized to the anisotropic one. However, the amount of damage that is generated depends on the sign of the direct stresses, i.e., tension or compression, so the damage growth equation has an inherent directional character even for the isotropic case.

A basic consideration in the constitutive equations of [2-6] is the decomposition of the total strain into reversible (elastic) and non-reversible (inelastic) components both of which are nonzero at all stages of loading and unloading, namely $\dot{\varepsilon}_{ij} = \dot{\varepsilon}^{e}_{ij} + \dot{\varepsilon}^{p}_{ij}$. For small strains, assumed here, $\dot{\varepsilon}^{e}_{ij}$ would be given by the time derivative of Hooke's Law. The governing equations for inelastic deformation are taken to be the Prandtl-Reuss flow law of classical plasticity,

$$\dot{\varepsilon}^p_{ij} = \dot{e}^p_{ij} = \lambda s_{ij} \qquad (1)$$

and a relation between the plastic strain rate and stress invariants that includes internal and external variables, $D^p_2 = D^p_2(J_2,Z,T)$. In these equations, $\dot{\varepsilon}^p_{ij}$ and $\dot{e}^p_{ij}$ are the direct and deviatoric plastic strain rates, s_{ij} the deviatoric stress, D^p_2 the second invariant of the plastic strain rate tensor, J_2 the second invariant of the deviatoric stress tensor, Z the scalar hardening parameter, and T the temperature.

Squaring (1) leads to $\lambda^2 = D^p_2/J_2$, so that, on the basis of (1) and the expression for D^p_2, the strain rate components could be expressed as functions of state quantities. A particular function was chosen for D^p_2 in the previous studies [2-6] which was motivated by equations proposed in the literature on "dislocation dynamics," namely,

$$D^p_2 = D^2_o \exp\{-(Z^2/3J_2)^n[(n+1)/n]\} \qquad (2)$$

where D^2_o is the asympotic value of D^p_2 and n is the material constant that controls strain rate sensitivity. Introducing a scalar damage state variable ω, in the same manner as Kachanov [1], into (2) leads to

$$D^p_2 = D^2_o \exp\{-[Z^2(1-\omega)^2/3J_2]^n[(n+1)/n]\} \qquad (3)$$

It seems reasonable to assume that the constants D_o and n and the evolution equation for the hardening parameter Z would be unaffected by the inclusion of ω. For the case of uniaxial stress, σ_x, the preceding equations would lead to the following expression for the axial plastic strain rate,

$$\dot{\varepsilon}^p_x = \frac{2}{\sqrt{3}}\left(\frac{\sigma_x}{|\sigma_x|}\right) D_o \exp\left\{-\frac{1}{2}\left[\frac{Z(1-\omega)}{\sigma_x}\right]^{2n}\left(\frac{n+1}{n}\right)\right\} \qquad (4)$$

It is noted that the basic flow law (1) is assumed to hold in the presence of damage, as was observed in [9].

In the earlier investigation, the evolution equation for Z was assumed to consist of a hardening term due to plastic work and a hardening recovery term, i.e.

$$\dot{Z} = \left(\frac{\partial Z}{\partial W_p}\right)\left(\frac{dW_p}{dt}\right) - A\left(\frac{Z-Z_i}{Z_1}\right)^r Z_1 \tag{5}$$

where A and r are temperature dependent material constants and Z_i and Z_1 are particular values of Z at a given temperature. A form suggested for $(\partial Z/\partial W_p)$ in [3,6] was $(\partial Z/\partial W_p) = m(Z_1-Z)$ where Z_1 is the upper limit on Z and m is a constant that controls the rate of hardening. The total expression for $\dot{Z}$ then takes the form,

$$\frac{\dot{Z}}{Z_1} = m\left(1-\frac{Z}{Z_1}\right)2\left(D_2^P J_2\right)^{1/2} - A\left(\frac{Z-Z_i}{Z_1}\right)^r \tag{6}$$

On the basis of (1), (2) and (6), the equations are able to predict primary and secondary creep and appropriate hardening response due to steady and changing strain rates for both loading and unloading conditions [3,6,8]. Use of (3), in place of (2), with the inclusion of the damage parameter, would enable the equations to also represent tertiary creep and other deformation characteristics close to the failure condition.

The growth equation for damage would be expected to be generally similar to (6) in that $\dot{\omega}$ should be a function of stress, the damage itself, and temperature. Direct coupling between hardening and damage, i.e., in their respective growth laws, is considered to be unimportant. Recovery of damage is not usually observed so a damage recovery term is not included. For creep under uniaxial stress, σ_x, a power law relation for $\dot{\omega}$ was proposed in [10] which was a generalization of the equation described in [1], namely,

$$\dot{\omega} = \left[\frac{\sigma_x}{\sigma_0(1-\omega)}\right]^s (1-\omega)^{-t} \tag{7}$$

where σ_0 and the exponents s and t are temperature dependent constants. This equation can be written more generally as

$$\dot{\omega} = f_1(\sigma_x) f_2(\omega) \tag{8}$$

which, in principle, can be integrated to give

$$f_3(\omega) = \int_t f_1(\alpha_x)dt \ . \tag{9}$$

Equation (9) is, in effect, a damage criterion with failure indicated by $\omega = 1$. Generalizing (9) to the multidimensional stress case could be achieved by replacing σ_x with suitable combinations of the stress tensor. Leckie and Hayhurst [11] suggest that a reasonable choice would be the combinations corresponding to the isochronous damage surfaces. On this basis, (9) would take the general form,

$$f_3(\omega) = \int_t f_1(\alpha\sigma_{max} + \beta\sqrt{3J_2} + \gamma I_1)dt \tag{10}$$

where I_1 is the first invariant of the stress tensor and α, β, and γ are constants ($\alpha + \beta + \gamma = 1$). In (10), σ_{max} is generally meaningful only when positive (tensile), i.e., $\sigma_{max} = \sigma^+_{max}$, and I_1 is also usually taken to be only positive (i.e., hydrostatic tension). A minimum requirement is that the integrand of (10) be positive, i.e., healing doesn't occur, but usage dictates that σ_{max} and I_1 take on only positive values.

Based on test results for isochronous failure surfaces, Leckie and Hayhurst [11] consider that f_1 could be expressed as

$$f_1(\sigma_{max}, J_2, I_1) = [\alpha\sigma^+_{max} + \beta\sqrt{3J_2} + \gamma I^+_1]^\nu \tag{11}$$

where the exponent ν is a material constant. Upon defining Q as

$$Q = \int_t [\alpha\sigma^+_{max} + \beta\sqrt{3J_2} + \gamma I^+_1]^\nu dt \ , \tag{12}$$

eq. (10) can be rewritten as

$$\omega = g(Q); \ \frac{dQ}{dt} \geq 0 \tag{13}$$

Eqs. (12) and (13) could be interpreted as an independent damage criterion which has the impulse form that has been suggested by some investigators, e.g., [12]. In the context of this formulation, (13) is inherently coupled to the strain rate response equations. The damage parameter ω is treated in (3) as an isotropic scalar, but its growth law has a directional character indicated by Q, (12).

The procedure usually followed in the literature is to assume forms for both functions f_1 and f_2 in (8), e.g., (7), and, by integration, to obtain an expression for $f_3(\omega)$, (9). This method could also be generalized to the multidimensional stress case as shown in [11].

An alternative procedure is to assume a functional form for f_1, e.g., (11), and also for g(Q), (13). Regarding g(Q), the exponential function given by (2) has the desired properties in that it is very small at low values of the independent variable, tends to increase rapidly beyond a transition region, and has an asymptotic value. For the case of damage, this could be expressed as

$$\omega = \exp[-(H/Q)^p] \tag{14}$$

where H and p are temperature-dependent constants to be determined empirically. The growth equation for ω would then be the time derivative of (14). A possible advantage of this procedure is that only the new material constants H and p need to be determined from special tests since the others, ν, α, β, γ, are obtained from isochronous damage surfaces. From the viewpoint of numerical computation, (14) is probably simpler to use than power law expressions.

It is noted from (14) that $\omega \to 1$ only as $Q \to \infty$. This does not mean, however, that failure requires an infinite time since (14) is coupled to the response equations and very large strains indicative of failure would occur in a finite time for most boundary value problems. Eq. (14) does avoid the need to introduce the specific rupture condition $\omega = 1$ when nonasymptotic functions of ω are obtained upon integrating the equations for $\dot{\omega}$. The possible conceptual and numerical advantages of this approach will be examined in subsequent studies.

B. Anisotropic Case

Induced damage appears to have both isotropic and anisotropic characteristics which influence subsequent deformations in a comparable manner. Damage is therefore a nonisotropic, history-dependent internal state variable similar to the work-hardening variable Z described in [5]. Damage has the additional complication in that the magnitude of the induced effects generally depends on the sign of the stresses at each stage in the loading history, i.e., it is greater for tension than for compression.

In order to formulate an initial theory for anisotropic damage, it will be assumed that the essential differences between tensile and compressive loadings could be accounted for in the reference stress parameter Q, (12). Since $\omega = g(Q)$, (13), the relative role of direct tensile stresses in causing damage would be indicated by appropriate values of α and γ in (12).

Considering ω in a manner similar to that of anisotropic hardening in [5] means that it is taken to be a second order tensor with a particular "cosine law" relation for the relative increases in the various components. Following [5], the damage increment $\Delta\omega$ due to an increment in the stress parameter Q, (12), would have an isotropic part $q'(\Delta\omega)$ and an anisotropic, directional part whose magnitude would be $(1-q')\Delta\omega$ in the current direction of the reference tensorial parameter (treated as a vector in a six-dimensional vector space). This reference should be the stress tensor for the case of damage (and also for hardening, although the analysis of [5] is based on the plastic strain rate as the reference).

The proposed growth law, from [5], states that the contribution of the anisotropic damage increment in all directions in the reference stress space would be the value $(1-q')\Delta\omega$ multiplied by the cosine of the angular direction relative to the current stress vector. Since the damage increment should be always positive, $q' \geq 1/2$, which means that at least half the incremental damage increase is isotropic.

The procedure for calculating the changes in the components of ω with loading history would then be identical to that given in [5] for Z, with Q, (12), replacing plastic work as the independent scalar variable and eq. (13) for $\omega = \omega(Q)$ replacing $Z = Z(W_p)$. The equation for plastic strain rate with the inclusion of the damage parameter, in the notation of [5], equation (5.5), would then become

$$D_\alpha = D_0 \exp\{-\frac{1}{2}\left[\frac{Z^2_{\alpha\alpha}(1-\omega_{\alpha\alpha})^2}{J_2}\right]^n\} \frac{T_\alpha}{\sqrt{J_2}} \tag{15}$$

where α here refers to the various coordinate directions in the determined six-dimensional stress space. D_α and T_α are the six-dimensional rate of plastic deformation and deviatoric stress vectors which are related to the ordinary plastic strain rate and stress tensorial components by a simple mapping. In (15), $Z_{\alpha\alpha}$ and $\omega_{\alpha\alpha}$ are the applicable values of the "hardening" and "damage" internal state variables corresponding to the indicated direction. All the components of the variables Z and ω are recomputed at each stage of the loading history according to the methods given in [5].

References

1 Kachanov, L.M.: On the Rupture Time under Creep Conditions, (in Russian). Izv. Akad. Nauk SSR, Otd. Tekhn. Nauk (1958) 26-31.

2 Bodner, S.R. and Partom, Y.: A Large Deformation Elastic-Viscoplastic Analysis of a Thick-Walled Spherical Shell. ASME J. Appl. Mech. 39 (1972) 751-757.

3 Bodner, S.R. and Partom, Y.: Constitutive Equations for Elastic-Viscoplastic Strain Hardening Materials. ASME J. Appl. Mech. 42 (1975) 358-389.

4 Bodner, S.R., Partom, I. and Partom, Y.: Uniaxial Cycling Loading of Elastic-Viscoplastic Materials. ASME J. Appl. Mech. 46 (1979) 805-810.

5 Stouffer, D.C. and Bodner, S.R., A Constitutive Model for the Deformation Induced Anisotropic Plastic Flow of Metals. Int. J. Engng. Sci. 17 (1979) 757-764.

6 Bodner, S.R.: Representation of Time Dependent Mechanical Behavior of René 95 by Constitutive Equations. AFML-TR-79-4116,(1979).

7 Derman, D., Zaphir, Z. and Bodner, S.R.: Nonlinear Anelastic Behavior of a Synthetic Rubber at Finite Strains. J. Rheology 22 (1978) 239-258.

8 Bodner, S.R. and Merzer, A.: Viscoplastic Constitutive Equations for Copper with Strain Rate History and Temperature Effects. ASME J. Engng. Mat. Tech. 100 (1978) 388-394.

9 Johnson, A.E., Henderson, J. and Khan, B.: Complex-Stress Creep, Relaxation and Fracture of Metallic Alloys. Edinburgh: H.M. Stationery Office (1962).

10 Rabotnov, Y.N.: Creep Problems in Structural Members. North-Holland Pub. Co. Amsterdam (1969).

11 Leckie, F.A. and Hayhurst, D.R.: Constitutive Equations for Creep Rupture. Acta Met. 25 (1977) 1059-1070.

12 Tuler, F.R. and Butcher, B.M.: A Criterion for the Time Dependence of Dynamic Fracture. Inter. J. Fracture Mech. 4 (1968) 431-437.

Acknowledgements

The author wishes to thank Professor F.A. Leckie for the benefit of his experience in many discussions on the subject of this paper. This work was performed under NSF Grant ENG-78-23549 with the University of Illinois, Urbana, Illinois, while the author was on leave from the Technion-Israel Institute of Technology, Haifa, Israel.

The Use of Statistical Reference Stress Techniques in Nonlinear Structural Analysis

J.T. BOYLE

Department of Mechanics of Materials
University of Strathclyde
Glasgow, Scotland, U.K.

Summary

It is the purpose of this paper to describe the use of the Reference Stress Method in structural analysis for material nonlinear behaviour as a simple means of assessing the effect of scatter in uniaxial test data. As an example the life to ductile rupture of a spinning holed disc is examined.

1. Introduction

Structural analysis for material nonlinear behaviour is well advanced: yet it can be argued that the sophistication of current techniques is far in advance of knowledge of actual material behaviour. A practical limitation imposed by this uncertainty is that observed scatter in simple testing assumes a new significance since it is amplified by the inherent nonlinearity of the material response. Therefore, unless this uncertainty can somehow be reduced, for example by statistical quantification, the usefulness of sophisticated solution techniques is questionable. One method of assessing the effect of scatter uses the concept of "reference stress" [1,2]. This allows the effect of material scatter on some characteristic response of a structure to be assessed from the results of a uniaxial test at the reference stress. Moreover, for some classes of structures it is often found that the reference stress is independent of the presumed material response, which has important implications for design.

2. Analysis

Although originally developed for metallic creep, the method can be used for any nonlinear material behaviour. Suppose that it is desired to examine some characteristic response of a structure, ρ, which may represent a strain, displacement, lifetime, etc. This is to be related to an analogous response of a uniaxial test, ρ_0,

$$\rho_0 = f(\sigma_0, p_1, p_2, \ldots) \tag{1}$$

where σ_0 is the stress at which the test is performed and $p_1, p_2, \ldots$ etc are the material parameters: there are no particular restrictions on the functional form of f.

By way of structural analysis the characteristic response can be expressed in the form

$$\rho = g(p_1, p_2, \ldots) \tag{2}$$

which can be related to the result of a uniaxial test

$$\rho = \delta \times \rho_0 \tag{3}$$

for some scaling factor

$$\delta(\sigma_0, p_1, p_2, \ldots) = g(p_1, p_2, \ldots)/f(\sigma_0, p_1, p_2, \ldots)$$

The aim of the reference stress method is to choose $\sigma_0 = \sigma_R$ such that δ is virtually independent of variations in the material parameters arising from scatter: this stress is called the reference stress and is chosen such that the statistical variance of the scaling factor vanishes

$$V[\delta(\sigma_R)] = 0 \tag{4}$$

In practice this must be done approximately and suitable methods have been discussed elsewhere by the writer [2].

3. An Example

As an example of the above, consider the problem of estimating the life to ductile rupture of a rotating disc in creep, described in detail by Kachanov [3]. The disc has thickness h initial inner and outer radii b_0 and a_0 with angular velocity ω; large strains are assumed and the maximum shear stress criterion is adopted as a simplification. It can be demonstrated that the normalised rupture time is given by

$$\tau_R = \int_0^{t_R} b(t)dt = \int_0^{\infty} F(U)dU \tag{5}$$

where for Norton's law $\dot{\varepsilon}_C = b(t)B_N\sigma^n$

$$F(U) = B_N \left[\frac{\dfrac{\beta^2}{\frac{1}{n}(1-\frac{1}{n})s} \left(\dfrac{\frac{\beta}{n}+U}{(\beta+U)^{1/n}} - \dfrac{\frac{1}{n}+U}{(1+U)^{1/n}} \right)}{\left(\frac{\beta^3-1}{3} + U\frac{\beta^2-1}{2}\right)} \right]^n$$

and for Dorn's law $\dot{\varepsilon}_c = b(t)B_D\exp(\alpha\sigma)$

$$F(U) = B_D\exp\left\{ -\left(\frac{\alpha s}{\beta^2}\left(\frac{\beta^3-1}{3} + U\frac{\beta^2-1}{2}\right) - \right.\right.$$

$$-(\beta-1-((\beta+U)\ln(\beta+U) - (1+U)\ln(1+U)))$$

$$-\frac{U}{2}(\ln^2(\beta+U) - \ln^2(1+U))$$

$$\left.\div(\beta-1-U\ln((\beta+U)/(1+U)))\right\}$$

where $\beta = b_o/a_o$ and $s = \gamma\omega^2 b_o^2$.

For a uniaxial test the lifetime to ductile rupture can be obtained as

$$\text{Norton's law} \quad \tau_{RO} = B_N/(n\sigma_o)^n \qquad (6)$$

$$\text{Dorn's law} \quad \tau_{RO} = B_D Ei(-\alpha\sigma_o)$$

where Ei is the exponential integral.

As a numerical example for 304SS(ORNL Heat 9T4296) the estimated mean ($\bar{\ }$) and variance ν are taken as [2]

$\bar{n} = 4.6875$ $\quad \nu_n = 0.1569$ $\quad \bar{\alpha} = 0.0002961$ $\quad \nu_\alpha = 0.000009$

then for the particular values

$\beta = 8$ $\quad s = 10000$ psi

the reference stress and scaling factor are obtained as

Norton's law $\sigma_R = 3809$ psi $\quad \delta = 0.54$

Dorn's law $\sigma_R = 3787$ psi $\quad \delta = 0.52$

It can be seen from this example that the reference stress and scaling factor are virtually independent of the material constitutive function used. Although this cannot always be guaranteed, it does have important implications for design.

References

1 Boyle, J.T.: Mech. Res. Comm. 4 (1977) 41-44.
2 Boyle, J.T.: Int. J. Mech. Sci. 22 (1980) 73-82
3 Kachanov, L.M.: The Theory of Creep. Transl. NLL (1967) Sect. 74.

Analyse d'une Structure Elastoplastique Soumise à un Chargement Cyclique

J. CASIER

Electricité de France. Service Etudes et Projets Thermiques et Nucléaires

B. BUFF

Université de Technologie de Compiègne. Département de Génie Mécanique

Résumé

Zarka et Casier ont proposé une méthode originale permettant d'obtenir directement des bornes de l'état limite d'une structure soumise à un chargement cyclique et ceci à partir de calculs élastiques. Cette méthode est appliquée ici à une structure soumise à un chargement de type non radial conduisant à l'adaptation. Les résultats sont comparés à ceux d'un calcul incrémental classique.

Introduction

La détermination de l'état limite d'une structure soumise à un chargement cyclique par un calcul incrémental conduit à des coûts très élevés. La méthode proposée en [1] et [2] ne faisant appel qu'à des calculs élastiques permet d'obtenir des bornes de cet état limite avec des temps de calcul réduits. Elle fut d'abord appliquée avec succès à des chargements radiaux, c'est-à-dire tels que la solution élastique en tout point de la structure varie linéairement en fonction du temps ([3] et [4]). Nous l'appliquons ici à un chargement non radial : la structure, une coque mince cylindrique, est soumise à un effort axial constant et à un gradient longitudinal cyclique. Le matériau est à écrouissage cinématique linéaire avec critère de Misès associé.

Principe de la méthode

Pour un tel matériau, le critère de plasticité s'écrit :

$$f\ (\sigma - \alpha) \leqslant 0 \tag{1}$$

où σ est le champ des contraintes et α le champ de paramètres internes dépendant des déformations plastiques.

Le principe de la méthode consiste à considérer σ comme la somme du champ des contraintes élastiques et d'un champ d'autocontraintes :

$$\sigma = \sigma^{el} + p \tag{2}$$

On associe ensuite aux paramètres α les paramètres $\hat{\alpha}$ par :

$$\hat{\alpha} = \alpha - p \tag{3}$$

le critère s'écrit alors :

$$f\ (\sigma^{el} - \hat{\alpha}) \leqslant 0 \tag{4}$$

Cette condition exprime que $\hat{\alpha}$ appartient à un convexe qui subit une translation lorsque le temps varie. Il y aura adaptation si l'intersection $\hat{C}_\ell$ de ces convexes est non vide. De plus, la valeur limite du champ $\hat{\alpha}$ doit appartenir à $\hat{C}_\ell$. Dans [2], Zarka et Casier proposent de la construire en projetant la valeur initiale de α sur $\hat{C}_\ell$. Dans le cas du chargement non radial, l'allure complexe du convexe $\hat{C}_\ell$ rend cette construction difficile. Nous proposons alors de remplacer C_ℓ par la plus grande boule qu'il contient et que l'on construit localement à partir de la solution élastique. Projetant sur cette boule la valeur initiale de α, obtenue par un calcul incrémental sur un demi-cycle, nous obtenons une borne supérieure dans l'espace des paramètres transformés de la solution du problème. L'existence d'une relation biunivoque entre α et $\hat{\alpha}$ permet alors d'accéder aux déformations plastiques [2].

Application :
La structure étudié et son chargement sont définis fig. 1. Le calcul incrémental mené sur 14 cycles avec le code PASTEL a nécessité 2h48mn de CPU. La déformation plastique évolue encore de 1% au cours du dernier cycle. L'application de la méthode a exigé 6mn30s de CPU dont 6mn pour le premier demi-cycle. Les profils de déformations plastiques le long de la paroi du tube sont comparés sur les fig. 2 et 3.

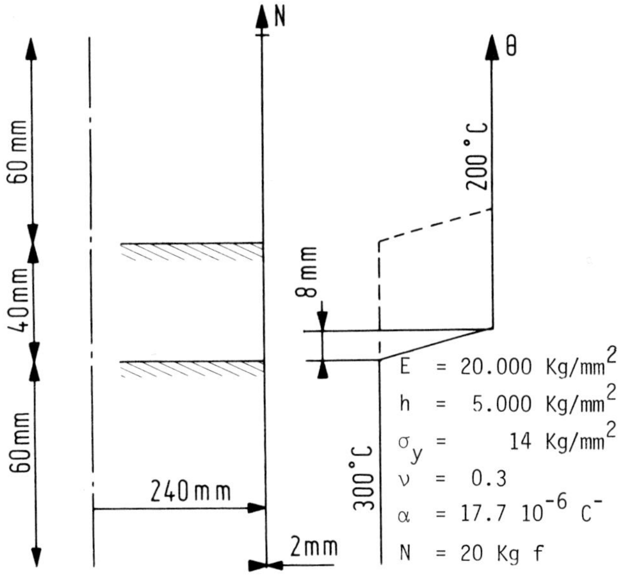

Fig. 1 - Structure et chargement étudiés

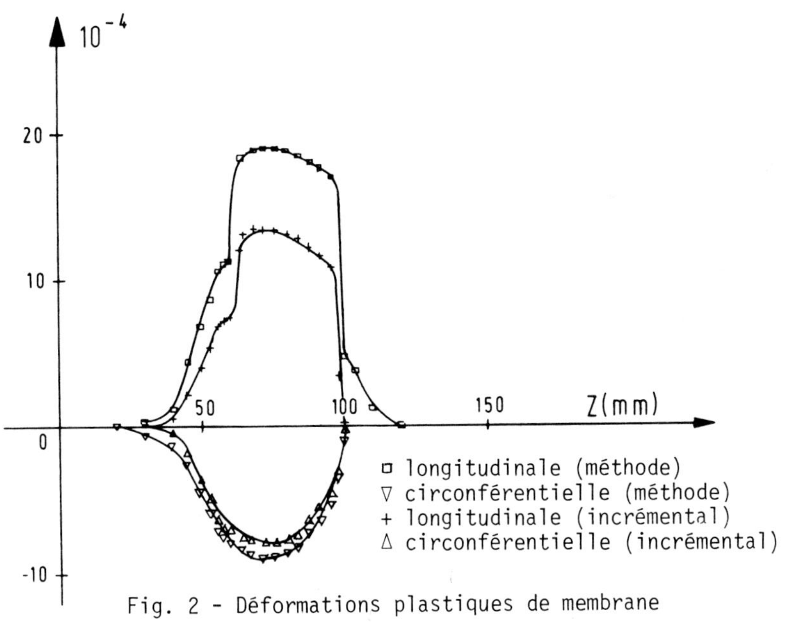

Fig. 2 - Déformations plastiques de membrane

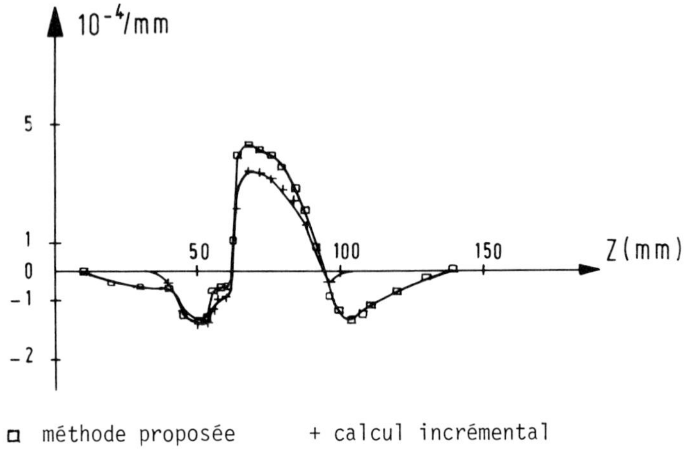

Fig. 3 - Déformation longitudinale de flexion

Références

1 CASIER, J. : Analyse d'une structure élastoplastique soumise à un chargement cyclique. Thèse de Docteur-Ingénieur présentée à l'I.N.S.T.N. le 23.02.78.

2 ZARKA, J. ; CASIER, J. : Cyclic loading on an elastic plastic structure. Practical rules. Mechanics Today, Nemat Nasser editor, à paraître.

3 EVENO, D. : Calcul de l'état limite d'une structure massive élastoplastique dans le cas de l'accomodation. EDF SEPTEM Note technique E-SE/TA-79-51.

4 CASIER, J. ; FAIDY, C. : Etude de coques axisymètriques sousmise à des chargements cycliques. EDF SEPTEM. Note technique E-SE/TA-79-13.

Discussion on Problems of Models Identification

J.L. CHABOCHE, J. LEMAITRE, D. MARQUIS, and S. SAVALLE.

O.N.E.R.A. Chatillon - sous - Bagneux FRANCE
LABORATOIRE DE MECANIQUE ET TECHNOLOGIE-Université PARIS VI-
E.N.S.E.T. Cachan. FRANCE.

Summary

The main difficulties arising in the problems of identification of numerical values for materials coefficients involved in strain or rupture constitutive equations are described with emphasis on highly non linear conditions. Direct identification can be used for simple models but computer assistance is needed when the number of unknown coefficients is greater than 3. Interactive method is described and some details are given on automatic minimization procedure of error functions, particularly on GAUSS-NEWTON method. Examples deal with plastic, viscoplastic and damage rupture behaviors.

1. Identification of mathematical models.

Mathematical models of materials behavior generally express phenomenological variables such as strain ε or plastic strain ε^p, stress σ, damage D, crack length ... etc... as functions or functionals of dependent variables and eventually internal variables α_j. The thermodynamical approach for example, gives rise to models made of a set of one, two, three or more simultaneous differential equations, often non linear, giving the rates of variables evolution as functions of actual values of all other variables [1] :

$$\dot{\varepsilon}^p = f(\sigma, \alpha_1 .. \alpha_j ..)$$
$$\dot{\alpha}_j = g_j(\sigma, \alpha_1 .. \alpha_j ..)$$

The analytical shape of functions f and g_j are chosen on the basis of physical and historical background, on the basis of qualitative results and we must say on the basis of applicability of the models and authors fancy ! (we love powers functions !). These functions involve coefficients defining the numerical evolutions, they are characteristic of each material : their values are different for different materials but constant for a given material.

To identify a model consists to find the values of these coefficients which fit the best a set of experimental results of specific tests. These results express the responses of the material to particular inputs which have to be fitted by the mathematical response of the model to the same inputs. Constitutive equations and response functions are of different nature : the first one represents the general material behavior within a frame of hypothesis, the second characterizes one particular case of behavior. Constitutive equations cannot be identified by themselves because only response functions are measurable.

The need for prediction of structure service behavior, the computers capability of complex calculations and the development of solid mechanics have given rise to complex models with many variables and by consequence many coefficients to identify especially when temperature acts as a variable.

Direct identification of two coefficients defining a highly non linear function is already a difficult problem to choose the specific tests and to obtain a good accuracy in a wide range of variables variation [2]. When the number of coefficients increases, interactive methods using computers to integrate the models and compare the numerical response to experimental results are necessary, but the uniqueness and the convergence are not warranted [3] . Automatic minimization procedures can be used. They give excellent results for linear or slightly non linear problems [4] , but remain very difficult to use for highly non linear problems [5] . What is the solution : automatic procedure with interactive judgement of the human operator ? use of artificial intelligence concepts ? This is an open problem.

2. Direct identification.

Just to precise the problems, let us see what is the usual direct identification on the example of perfect visco-plastic NORTON's relation [6] :

$$\dot{\varepsilon}_p = \left[\frac{\sigma}{\lambda}\right]^N$$

where λ and N are the two coefficients to be determined.

- If all the state variables are measurable : here the stress σ and the plastic strain $\varepsilon_p = \varepsilon - \frac{\sigma}{E}$ can be measured as a function of time t (E is the YOUNG modulus).
- If response functions of the model are explicit : here creep response is $\dot{\varepsilon}_p = \left[\frac{\sigma}{\lambda}\right]^N$, relaxation response is $\frac{\dot{\sigma}}{E} + \left[\frac{\sigma}{\lambda}\right]^N = 0$
- If coefficients are separable in the response function : here $\text{Log}\ \dot{\varepsilon}_p = -N\,\text{Log}\,\lambda + N\,\text{Log}\,\sigma$ for creep
  $\text{Log}\left(-\frac{\dot{\sigma}}{E}\right) = -N\,\text{Log}\,\lambda + N\,\text{Log}\,\sigma$ for relaxation.

Then, direct identification is applicable. Here a plot of $\dot{\varepsilon}_p, \sigma$ or $\left(-\frac{\dot{\sigma}}{E}\right)$ from test measurements on Log-Log co-ordinates gives N and then λ .

In order to obtain a good accuracy with a minimum number of tests, the specific tests have to be chosen carefully. In particular all the variables have to be excited in a range of variation wide enough. Here one (secondary) creep test gives only one identification point ($\dot{\varepsilon}_p, \sigma$) which is not sufficient to identify the two coefficients λ and N . Several tests, taking a long time, are needed. A better procedure consists in using one relaxation test in which the stress σ varies in a wide range of stress rate $\dot{\sigma}$ [7] .

If one of these three restrictive conditons does not exist in models, the direct identification is difficult to apply and there is a need for computer assistance.

3. Interactive method of identification.

Many useful models involve internal variables which are not directly measurable :

- plasticity or visco-plasticity with kinematic and isotropic variables [8] [9] [10] .
- ductile, creep or fatigue rupture with damage variable [11] [12] [13] [14] .
- crack propagation with memory effect threshold variable [15] .

These internal variables are only indirectly measurable by their effects on observable quantities such as stress, strain, rupture time, crack length. This means that the working quantities, with which can deal the identification, are the evolutions of the observable variables for some responses to specific inputs ; such responses being not explicit, the identification by interactive procedure corresponds to the following steps :

a) introduce test results (response functions) and corresponding inputs in the computer,
b) choose initial set of coefficients
c) calculate the response of the model equations to the particular inputs for the set of coefficients,
d) observe the difference between experimental and calculated response, think hard !
e) if this difference is not admissible (larger that the test discrepancy) choose a new set of coefficients and go back to step c.

This procedure is greatly accelerated when using graphical computer terminal, but it needs a good qualification of operator : (dexterity, experience) and the result may be somewhat subjective ! Furthermore it is almost impossible to practice as the number of unknown coefficients become greater than 3.

4. Automatic minimization methods.

Except the choice of initial values of coefficients, the whole former method can be set automatic using an error function to compare calculated and measured responses and minimization technics to find better sets of coefficients.

Error function

The mean square difference is the most used scalar norm but it can yield difficulties if it is used on the simple difference of variables for such non linear functions as the one shown on Fig. 1. The two values :

$$V = \sum_i v_i^2 = \sum_i (Y_i - Y_{i\,exp})^2$$

or

$$U = \sum_i u_i^2 = \sum_i (X_i - X_{i\,exp})^2$$

do not give the same result and give a different importance to the different parts of the curve.

The mean square "distance" avoids these difficulties and has been used with succes [16].

$$D = \sum_i d_i^2 \simeq \sum_i \frac{u_i . v_i}{\sqrt{u_i^2 + v_i^2}}$$

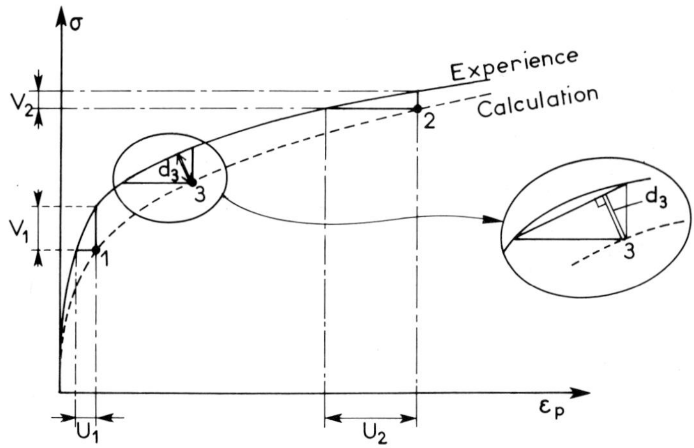

Fig. 1 Mean square distances on the monotonic hardening curves.

Automatic minimization procedures

Once the error function has been calculated for a set of coefficients, minimization procedures can be used in order to find a new set of coefficients closer to the solution :

- direct search methods : FIBONACCI [17] , ROSENBROCK [18] , SIMPLEXE [19]. No derivatives are involved but the convergence is slow for highly non linear problems and uniqueness is not warranted for narrow valleys.
- gradient methods : steepest descent [20] , conjugate gradient[4]. They use first derivatives but the convergence is still slow for narrow and curved valleys .
- NEWTON's method [4][21], it uses the second order approximation of the error function which involves second derivatives difficult to obtain with accuracy. The convergence is bad for materials behavior models and divergence may occur in non convex zones.
- GAUSS-NEWTON method [4] : it is a first order approximation of the NEWTON's method, which works only for mean square functions. It appears that it is the best (or the least weak !) method

for the type of problems discussed here because, when it converges, it converges rapidly ; when it diverges, this is due to a bad choice of initial set of coefficients and as this divergence is also rapid, it allows to change these coefficients without consuming too large computer time.

Principal difficulties.

No one of these methods gives a rigorous and unique result, due to the following characteristics of almost all problems of material behavior models identification.

- High non linearity of the problem which is due to many causes :
. representation of the physical non linearities,
. experimental discrepancy within one test or between several tests,
. bad validity of the model in regard to tests results,
. bad sensibilization of some unknown coefficients if the specific tests do not excitate all the variables.
- Non uniqueness of minimum if it exists :
. incompatibility between tests,
. incompatibility between model and tests results,
. insufficient experimental informations,
- Bad shape of the minimum : several typical characteristics are shown on the Figures 2 and 3.

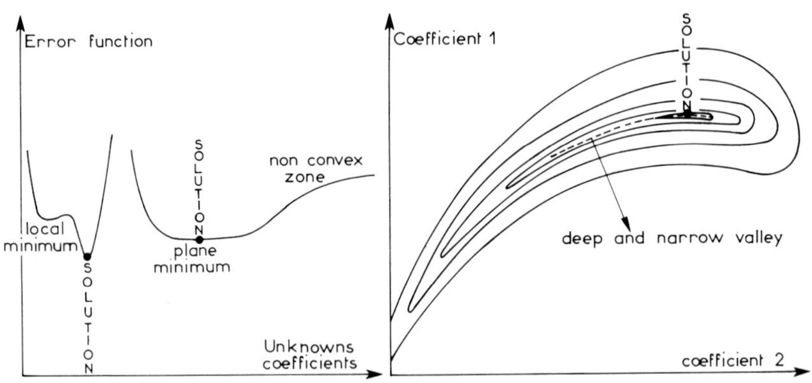

Fig. 2   Typical minima dans iso-error curves

Applications of GAUSS-NEWTON method.

Two examples will show the applicability of GAUSS-NEWTON method :

- The first one is the identification, for 316 L stainless steel at room temperature, of the three coefficients σ_0 , K and m occuring in the simple monotonic plastic relation :

$$\sigma = \sigma_0 + K\varepsilon_p^m$$

The experimental hardening curve shown on Figure 3 is given by 90 points. The function to minimize is the following :

$$V = \sum_i v_i^2 = \sum_i (\sigma_i - \sigma_0 - K\varepsilon_{p_i}^m)^2$$

which leads to the following system to be solved

$$\left(\sum_i \left\{\frac{\partial v_i}{\partial p}\right\}^T \left\{\frac{\partial v_i}{\partial p}\right\}\right) \{\Delta p\} = - \sum_i \left\{v_i \frac{\partial v_i}{\partial p}\right\}$$

T is the transposition operator

$\{\Delta p\}$ is the matrix column of σ_0 , K , m coefficients increments. Figure 3 shows the convergence of the GAUSS-NEWTON procedure for the following initial set of values : $\sigma_{01} = 100MPa$ $K_1 = 7500MPa, m_1 = .7$: Eight iteractions taking some seconds of computer give the good result.

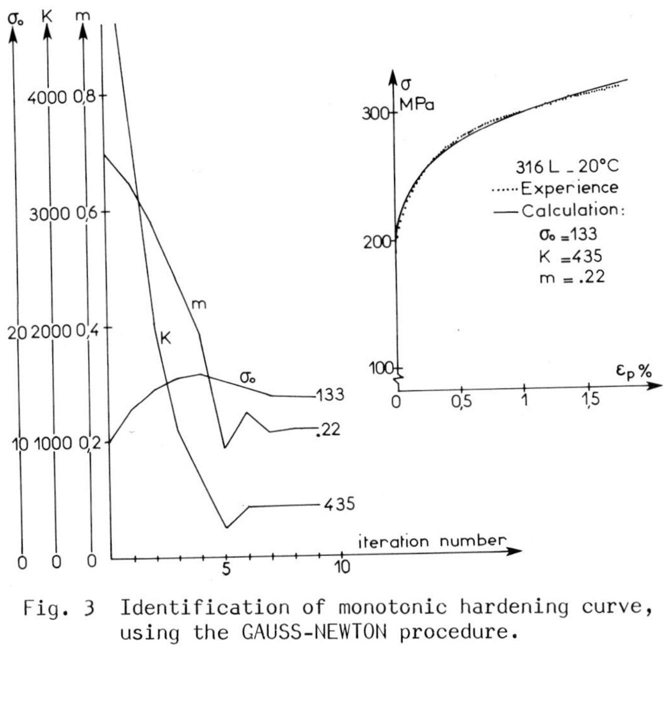

Fig. 3 Identification of monotonic hardening curve, using the GAUSS-NEWTON procedure.

- The second example is the identification of the stabilized cyclic plasticity model developped by MARQUIS [22]:

  The plastic loading function is :

$$f = J(\sigma - X) - R = \sqrt{\tfrac{3}{2}(\sigma' - X')(\sigma' - X')} - R \leqslant 0$$

  σ' and X' are the deviators of σ and X

  X is the internal stress tensor of kinematic hardening

  R is the width of the yield surface constant at stabilization.

  The normality rule is applied to the plastic strain rate :

$$\dot{\varepsilon}^p = \dot{p}\frac{\partial f}{\partial \sigma} \qquad \text{with} \qquad \dot{p} = \sqrt{\tfrac{2}{3}\dot{\varepsilon}^p : \dot{\varepsilon}^p}$$

  but is not applied to the variable associated to X :

$$\dot{X} = \tfrac{2}{3} C \dot{\varepsilon}^p - [\phi_\infty + [\phi_0 - \phi_\infty] \operatorname{Exp} - \gamma p] X \dot{p}$$

  5 coefficients R , C , ϕ_∞ , ϕ_0 , γ describe completely the monotonic and cyclic plastic behavior of each material :

  - initial hardening and BAUSCHINGER effect
  - cyclic stress strain loops
  - cyclic hardening or softening and stabilization.

Let us see the development of the method applied to the plastic cyclic behavior of 316 L stainless steel .

The stabilized cyclic behavior is first identified, under such conditions the kinematic equation becomes :

$$\dot{X} = \tfrac{2}{3} C \dot{\varepsilon}^p - \phi_\infty X \dot{p}$$

which leads to explicit integration under tension compression loading . In that case three coeffficients only are concerved and the stress is :

$$\sigma = R + \nu \frac{C}{\phi_\infty} + [X_0 - \nu \frac{C}{\phi_\infty}] \operatorname{Exp} - \phi_\infty \nu (\varepsilon_p - \varepsilon_{p_0})$$

where $\nu = \operatorname{Sign}(\dot{\varepsilon}_p)$ is the index of tensile or compressive flow and X_0 , ε_{p_0} are the values of X and ε_p at the beginning of each reversal. The GAUSS-NEWTON procedure has been applied independently for two particular test results [22] , giving rise to two sets of coefficients R , ϕ_∞ , C . Figure 4 shows the comparisons between experimental loops and calculated ones with mean values of coefficients : $R = 200MPa$, $\phi_\infty = 365$, $C = 72350$ MPa. Another procedure would be the introduction of the two experimental loops in only one minimization.

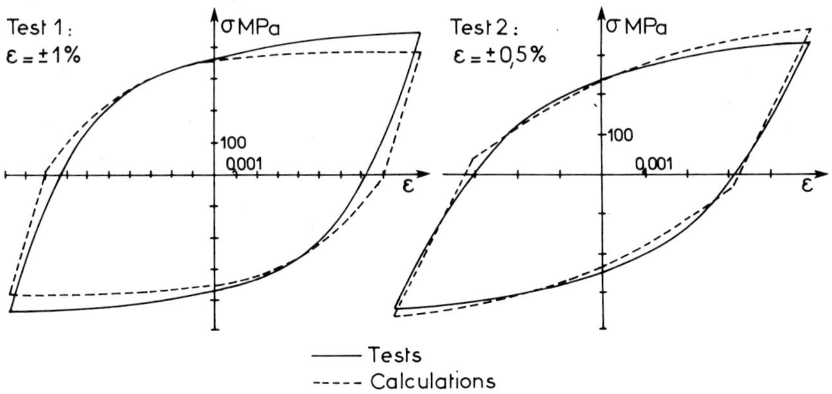

Fig. 4 Comparison between experimental and calculated stabilized stress-strain loops of 316 L stainless steel at room temperature.

After identifying the stabilized conditions, the initial behavior including BAUSCHINGER effect, and progressive cyclic hardening are described by the two last coefficients $\gamma = 11.2$ and $\phi_o = 623$. This shows how the automatic procedures can be used with successive steps, when the response functions are separated for special loading conditions.

The validity of a given representation of material behavior needs some verifications by comparing experimental and calculated responses for different loading conditions from those used to identify the coefficients. Figure 5 gives an example of such verification for cyclic stress controlled loading [22].

.../...

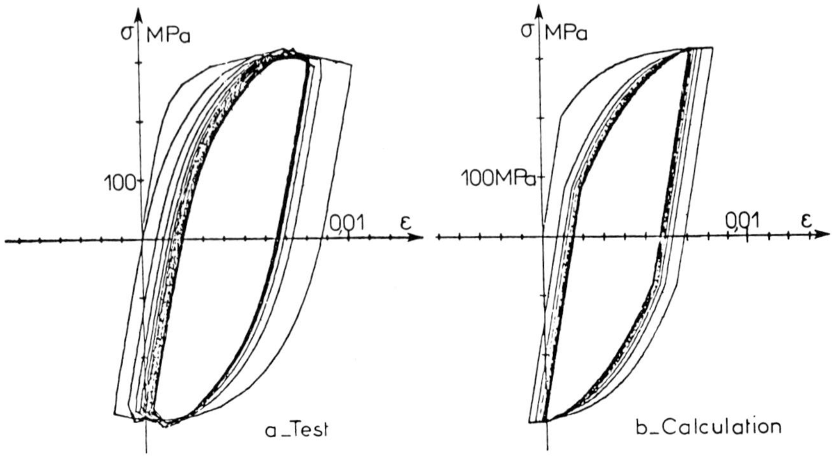

Fig. 5 Cyclic hardening under completely reversed stress control : a) test results , b) numerical simulation.

Improvements of the GAUSS-NEXTON method.

For situations where the behavior is more complex and when more coefficients have to be identified, application of the GAUSS-NEWTON procedure needs the development of particular strategies. Three kinds of procedures are briefly proposed and illustrated on the example of cyclic viscoplastic equations, applied to the IN 100 refractory alloy at temperature of 1000°C. Under stabilized cyclic tensile compressive conditions the constitutive equations used[23] [24] lead to similar relations as for the plasticity model studied above :

$$\sigma = R + \nu \frac{c}{\phi_\infty} + \left[X_0 - \nu \frac{c}{\phi_\infty} \right] \mathrm{Exp} - \nu \phi_\infty (\varepsilon_p - \varepsilon_{p_0}) + K \, |\dot{\varepsilon}_p|^{\frac{1}{n}}$$

where the additional viscous term introduces two more unknown coefficients. These particular operations are :

a) Reduction of test scattering by some pre-scaling : this is necessary because many different tests are needed to identify the stabilized cyclic behavior, especially tests including hold times or various frequencies (in order to sensibilize the rate effects). Only a few stabilized loops (here five), are introduced in the automatic minimization procedure but they are representative of more than twenty test results, because of the pre-scaling of plastic strain and plastic strain range.

b) Blockage of some coefficients to specific values, which is a way to introduce more informations than contained in the chosen test results : for example, the rate effect can be insufficiently sensibilized, but blocking the value of coefficient n to 7 results from :

- experience in the IN 100 behavior obtained from the litterature and the secondary creep rate equation (NORTON's equation indicated above),
- adjustments between the values obtained at several temperature levels : it has to be approximated by a smooth function of temperature.

This blockage procedure could also be useful to introduce some qualitative informations and constitutes a good way to improve the convergence of the minimization procedure and reduce the number of unknown coefficients.

c) Use of one particular unknown coefficient as a scaling parameter. In the highly non linear constitutive equations it is better to incorporate a particular coupling between the unknowns. Figure 6 schematically illustrates this effect for the tensile test equation used in section " Applications of GAUSS-NEWTON method", with $\sigma_0 = 0$. A change in value of exponent m (from 0.1 to 0.3), that is an adjustment of the non linearity of the model, could be interpreted as a bad change, although a simultaneous change of coefficient K (the scaling operation) leads to interpret the new value of exponent m as a better one.

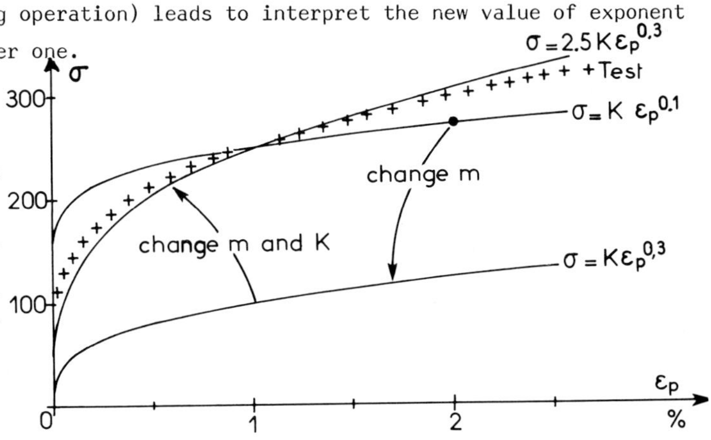

Fig. 6 Schematic illustration of the use of K as a scaling parameter (hypothetical monotonic curve)

In each constitutive equation there is at least, one particular coefficient like K, which corresponds to a natural scaling (giving for example the unity of the main variable : here the stress). Hence the following special procedure could be used to minimize the error functions which depends of I unknowns :

$$V = V(P_i, K) \qquad i = 1, \ldots, I-1$$

For each iterative step of the GAUSS-NEWTON method, made with the I-1 unknowns p_i one incorporates a particular minimization of V in regard to coefficient K (with p_i considered as constant) :

$$\tilde{K} = \tilde{K}(p_i) \quad \text{minimizes} \quad \tilde{V}(K) = V(p_i, K)$$

The minimization of V leads to a problem with I-1 unknowns (the last one being adjusted independently) :

$$p_i^* \text{ minimizes} \quad V^*(p_i) = V(p_i, \tilde{K}(p_i)) \qquad i = 1, \ldots I-1$$

Application of this concept is particularly interesting when the one dimensional minimization in regard to K can be made explicitely. In such cases, the proposed method saves computer time and improves efficiency of algorithm. Ithas been applied with success for the IN 100 case, by controlling a particular point (σ_E , ε_{pE} , $\ddot{\varepsilon}_{pE}$) in the viscoplastic equation, giving rise to :

$$K = \left[\sigma_E - R - \nu\frac{c}{\phi_\infty} - \left(X_o - \nu\frac{c}{\phi_\infty}\right) \mathrm{Exp} - \nu\phi_\infty[\varepsilon_{p_E} - \varepsilon_{p_o}]\right] \mathring{\varepsilon}_{pE}^{-\frac{1}{n}}$$

In this c ase, the minimization procedure was used in two steps : first with I-1 unknowns for this particular method and, after convergence, with I coefficients, opening this linkage.

Incorporating the three kinds of particular procedures, into the GAUSS-NEWTON method has lead to a correct identification of the IN 100 cyclic behavior, with a sufficient convergence for different sets of initial coefficients : Figure 7 indicates the calculated and experimental stabilized stress-plastic strain loops with the best solution : n = 7, K = 820, R = 25 MPa , C = 172250 MPa, ϕ_∞ = 650.

.../...

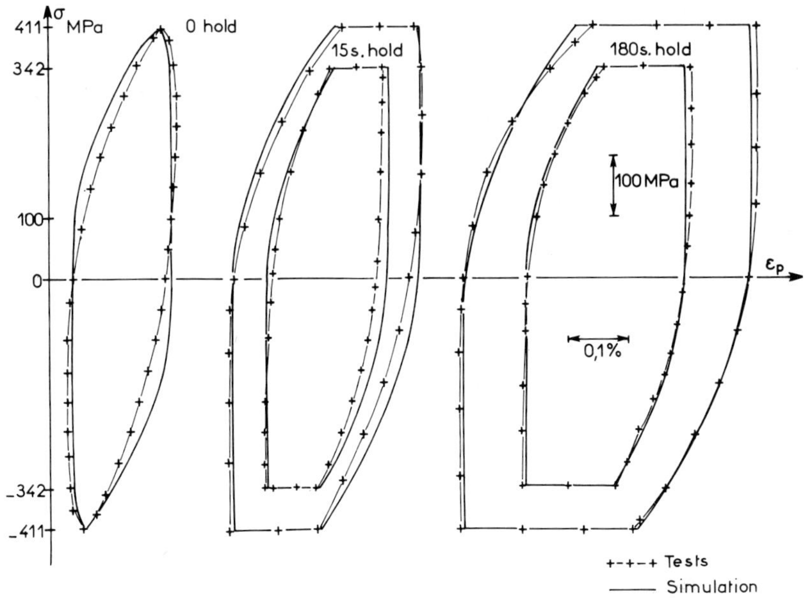

Fig. 7 Comparison between experimental and calculated stabilized stress-strain loopq od IN 100 alloy at 900°C.

5. Conclusion

What is the best procedure to identify mathematical models of behavior of materials ? No unique answer can be given. The following remarcks taken from some years of experiences can help but they cannot set up the final philosophy.

- Automatic and interactive procedures have to be coupled. The human operator with all his knowledge must interfer during the automatic procedure to evaluate the results and change the importance of one test or another or to change initial values.
- The experimental results must be treated to reduce discrepancy by smoothing procedure before introduction in the automatic procedure.
- The GAUSS-NEWTON method seems to be the most appropriate method for materials behavior modelling.
- Coefficients must be determined using a good choice of experimental tests.
- Block coefficients which are not sensibilized enough by tests results.
- Use one special coefficient as a scaling factor.

What improvement for the future ? It seems that artificial intelligence could help in the automatic procedure to choose algorithms or initial values that have given good results in previous similar problems in the past.

But remember three things :

a) Phenomenological models must be checked on experimental results of special tests that have not be used for the identification, especially multiaxial tests.

b) The "a priori" validity range of phenomenological models is the one that has been involved for each variable in the experimental tests used for identification.

c) Many physical knowledge or qualitative informations cannot be quantified and the last choice has to stay under the decision of the scientists.

BIBLIOGRAPHY

1 Kratochvil, J. : Thermodynamics of elastic-plastic materials, "Dynamique des Matériaux Inélastiques, Publ. Polish Acad. of Sciences, P. PERZYNA, 1974.

2 Lemaitre, J. : Elasto-visco-plastic constitutive equations for quasi-static structure calculations "Mechanics of inelastic materials , Publ. Polish Acad. of Sciences, 1972.

3 Himmelblau D.M. : Applied non linear programming, Mc Graw Hill ed, 1971.

4 Broyden, C.G; Fletcher, R.; Powell, M.J.D.; Swann, W.H.; Murray, W. : Numerical methods for unconstrained optimization, ed. Murray, W., 1972.

5 Galharague, C. : Applicabilité des méthodes d'identifications aux problèmes non linéaires. Rapport interne Laboratoire de Mécanique et Technologie, Université PARIS VI - ENSET, 1979.

6 Odqvist, F.K.; Hult, J. : Kriechfestigkeit metallischer, Springer-Verlag, 1962.

7 Lemaitre, J. ; Baudin, G. : Réalisation et utilisation des essais de relaxation pour l'étude du fluage de longue durée des métaux, 6ème Congrès International de Rhéologie, Lyon, 1972.

8 Rice, J.R. : On the structure of stress-strain relations for time-dependent plastic deformation in metals, J. Applied Mech. Trans. ASME, 37, 1970, p. 728-737.

9 Sidoroff, F. : On the formulation of plasticity and viscoplasticity with internal variables, Arch. Mech, vol 27, 5-6, 1975, p. 807-810.

10 Malinin, N.N, Khadjinsky, G.M. : Theory of creep with anisotropic hardening, Int. J. Mech. Sci., vol 14, 1972.

11 Janson, J,; Hult, J. : Fracture Mechanics and Damage Mechanics, A combined approach, J. Méc. Appliquée, Vol 1, 1977.

12 Krempl, E. : On phenomenological failure laws for metals under repeated and sustained loading (fatigue and creep). Conf. on environmental degradation of engineering materials, Blacksburg Virginia, 1977.

13 Lemaitre, J.; Chaboche, J.L. : Aspect phénoménologique de la rupture par endommagement, J. de Mécanique Appliquée, vol 2, N° 3, 1978, p. 317-365.

14 Chaboche, J.L. : The continuous damage mechanics : a tool to describe phenomena before crack initiation, SMIRT 5, Post Conf. Berlin, 1979.

15 Pellas, J.; Baudin, G.; Robert, M. : Fatigue crack growth model with two coupled differential equations, fracture 1977, ICF 4, Waterloo, Canada, Vol/2, p. 1353-1360.

16 Lemaitre, J.: Monthulet, A.: Recourcé, J.C. : Evaluation du facteur d'intensité des contraintes par mesures directes sur la structure fissurée chargée. La Recherc. Aérospatiale, N° 1974-5.

17 Fletcher, R. : Function minimization without evaluating derivatives. a revieuw. Comput. J. 8, 33-41. 1965.

18 Rosenbrock, H.H. : An automatic method for finding the greatest or least value of a function. Comput. J.3, 175-184.1960.

19 Nelder, J.A. ; Mead, R. : A simplex method for function minimization. Comput. J.7, 308-313. 1965.

20 Zoutendijk, G. : In J. Abadie (ed) "Non-linear programming-computational methods, in integer and non-linear programming." North-Holland, Amsterdam. 1970.

21 Greenstadt, J.L. : On the relative inefficiencies of gradient methods. Maths Comput. 21, 360-367. 1967.

22 Marquis, D. : Modélisation et identification de l'écrouissage anisotrope des métaux, Thèse ENSET-Université PARIS VI, 1979.

23 Chaboche, J.L. : Viscoplastic constitutive equations for the description of cyclic and anisotropic behavior of metals, Bulletin de l'Acad. Polonaise des Sciences. Vol. 25, N° 1 1977.

Strain Hardening Effects and Damage in Plastic Fatigue

G. CORDIER - K. DANG VAN

Laboratoire de Mécanique des solides de l'Ecole Polytechnique
91128-Palaiseau Cedex, FRANCE

Summary

A physical approach is presented. It takes into account the effect on the behaviour of the material of both strain hardening and rupture of the constitutive grains on one hand and of average value on the other hand. The model unables us to describe the evolution of the fatigue life with the applied loads and also the effects of temporary over or underloading.

Most of the recent studies on plastic fatigue are based on the concept of damage [1,2]. In Lemaitre and Chaboche theory, damage (defined by means of the effective stress) is supposed to be isotropic and characterized by a scalar D. D=0 in the virgin specimen and D=1 when a macrocrack is detected. However the interaction between strain hardening and damage has not been taken into account and it is not possible to describe by this theory the increase of fatigue life due to prior loading as it is observed.

In our model, macroscopic crack is created by junction of microscopic cracks on the heavy slip bands of the grains when the sound part of the matrix cannot support the load anymore.

In the following we note by Σ, E(resp. σ , ε) the quantities corresponding to the matrix (resp. to the grain).

We make the following assumptions:

H1-The grains as well as the polycristal are elastically isotropic.

H2-The orientations of monocristalline grains are randomly distributed.

H3-For each monocristal only one slip system S corresponding to the maximum effective shear stress τmax can be active. Each system S is defined by n_S and m_S normal and direction of slip (Fig.1). When slip has occured

only S will be active for further loading.
H4- For sake of simplicity we choose the interaction model of Sachs.
From H1 and H3 we have

$$E^p_{ij} = \frac{1}{V}\int \varepsilon^p_{ij}\, dV = \frac{1}{V}\int \alpha_{ij}\gamma\, dV$$

where γ is the slip and α the orientation matix of the grain defined by

$$\alpha_{ij} = \frac{1}{2}\left(n^S_i m^S_j + n^S_j m^S_i \right)$$

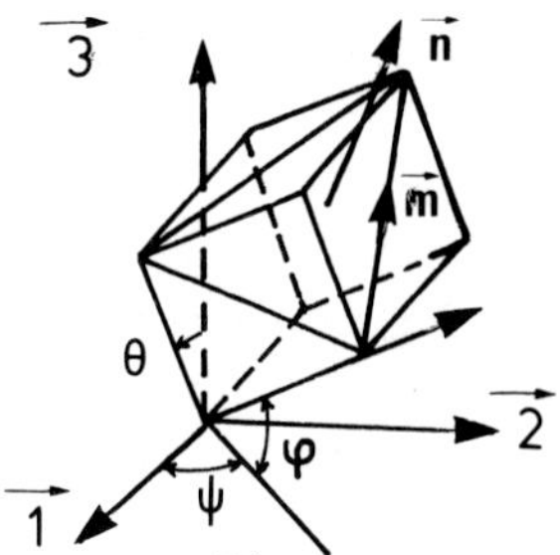

Fig. I

If a traction $\Sigma 3$ is applied, we have (from $\alpha_{ij}\alpha_{ij} = \frac{1}{2}$, $\alpha_{ii} = 0$)

$$\tau = \alpha_{33}(\psi,\theta,\phi)\Sigma$$

One can see that for each grain the principal slip system is defined by the maximum of its orientation factors. For c.f.c. material Zaoui [3] has shown that α^{max}_{33} can only take value between $2/3\sqrt{6}$ and 0.5. More precisely, τ_o being the initial slip threshold of grain, if one applies Σ to the virgin specimen with $2\tau_o < \Sigma < 3\sqrt{6}/2\tau_o$, then it exists a value ν function of Σ greater than $2/3\sqrt{6}$ such that only the grains defined by $.5 > \alpha_{33} > \nu$ can slip. Zaoui has also computed the normalized probability of α_{33} noted by $P(\alpha_{33})$.
Strain hardening and fracture laws are supposed to be identical for all the grains; in particular if $\tau > \tau_R$ the grain breaks; τ_R is a decreasing function of hydrostatic tension. For alternated cyclic tension, Σ is related to ΔE^p.
At a given time t, there are in the material three classes of grains which are characterized by their orientation factor

$2/3\sqrt{6} < \alpha_{33} < \lambda(t)$ the grains do not slip
$\lambda(t) < \alpha_{33} < \mu(t)$ the grains slip but are not broken
$\mu(t) < \alpha_{33} < .5$ the grains are broken

<u>Damage</u>

At first sight, one can define D by

$$D = \int_{\mu(t)}^{.5} P(\alpha_{33})\, d\alpha_{33} \quad \text{and when } \mu = 2/3\sqrt{6},\ D = 1$$

In fact a more precise definition is necessary. The obtained results show that at a given time of the damage process, the number of unbroken grains becomes insufficient to balance the imposed loading. We consider that at that time the volume is broken and D =1. In what follows, D is defined as the purcentage of broken grains relative to the total number of broken grains when the macroscopic crack is created.By that definition D depends not only of the past history but also of the loading at t_+. To follow the evolution of the damage of the material the system of non linear equations to be solved step by step is:

$$\tau = \alpha_{33}\,\Sigma \qquad , \qquad E^p = \int_{\lambda}^{\mu} \gamma\alpha_{33} P(\alpha_{33})\, d\alpha_{33}$$

$[\tau] = [\tau^i] + c.[\Delta\gamma]$ for a grain plastically deformed

$[\tau] = \tau_o$ for a grain being active for the first time

$\lambda = \mathrm{Inf}\ (\ \lambda' | \lambda'\Sigma > \tau_o ; 0.5\ )\quad , \quad \mu = \mathrm{Inf}\ (\mu' | \mu'\Sigma > \tau_R\ ;\ 0.5\ )$

Results

Fig.2 shows results we obtain in one level symetric traction compression. Dotted points are experimental results for stainless steel. The chosen value of the parameters are:

$$\tau_o / c = 400$$

$$\tau_R = \tau_o (1 + 0.5\ e^{-100\Delta E^p})$$

For $.6\% < \Delta E^p < 2\%$,we find exactly the experimental curve. For $\Delta E^p < .6\%$, correlation is bad because Sachs's model gives good results only for relatively important deformation.

Fig.3 shows the evolution of D versus N/N_R for different ΔE^p. Simulation of two levels loading tests is represented on fig.4. Curve 1 (resp. 2) shows the evolution of D after 480

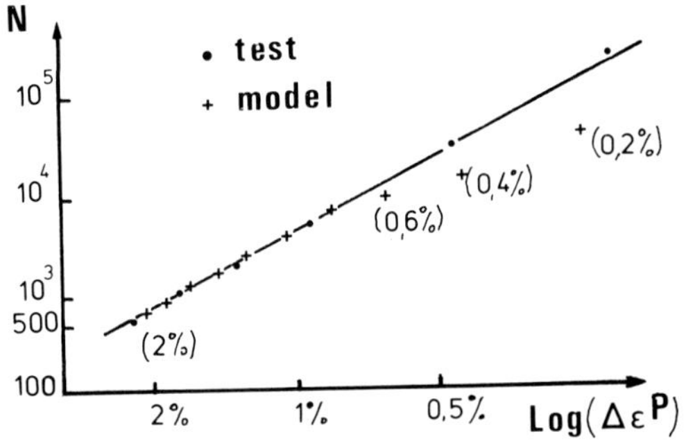

Fig. 2

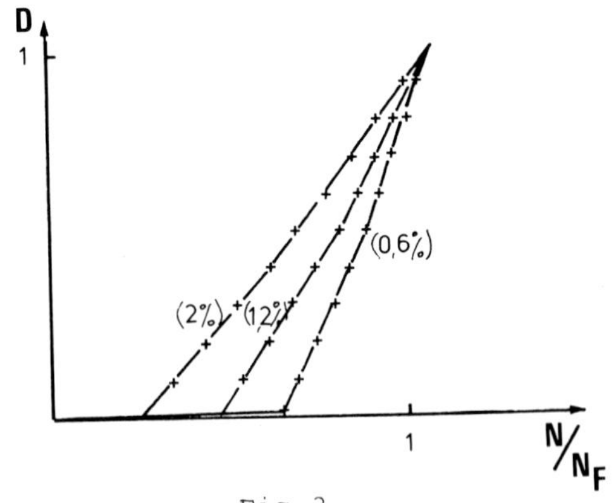

Fig.3

cycles at .6% (resp. after 640 cyles at.6%). These two curves are very different from the one level damage curve at ΔE^p = 2% (curve 3). Thus, the curve (D,N/N_R) depends not only on the past but also on the further loadings : this aspect seens to be surprising. In fact it comes from choosing D = 1 when rupture occurs.

Fig. 5 represents the cumulation curves for two levels loadings. In case of a high amplitude followed by lower amplitude loading,we obtain a curve largely below Miner's straight line. If the sequence is inversed, the results are more complex. Nevertheless, we find the classical trends of increase in fatigue life. It should be noted that N_2/N_R can be greater than 1. This effect has been observed experimentaly but cannot be describe before by any existing model.

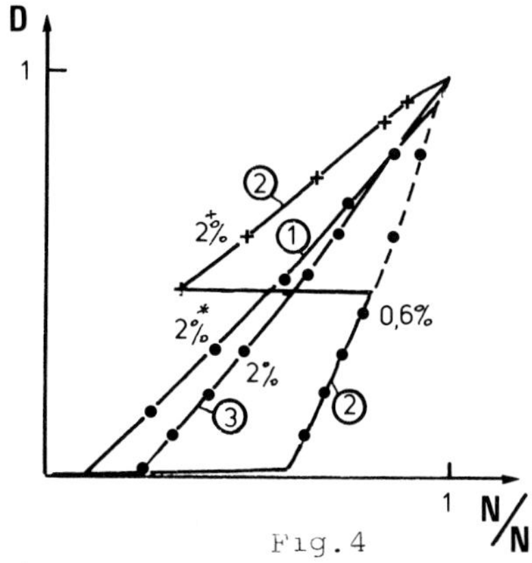

Fig.4

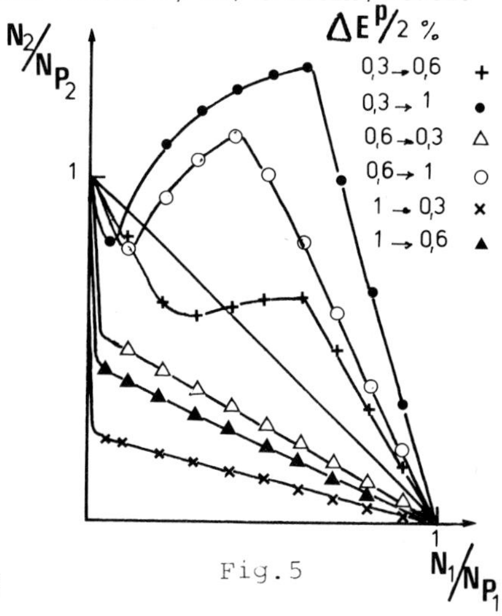

Fig.5

Literature

1 Bui-Quoc, T., "Dommage cumulatif en fatigue", Fatigue des Matériaux et des Structures, Masson, Paris, 1980.

2 Lemaitre, J. ; Chaboche, J.L. : Aspect phénomènologique de la rupture par endommagement, Journ. de Mécanique Appliquée, III, 3, 1978.

3 ZAOUI, A. : Etude de l'Influence Propre de la Désorientation des Grains sur le comportement Viscoplastique de Métaux Polycristallins (Système C.F.C.), Thèse de Doctorat ès Sciences, Paris,1970.

The Concept and Application of the Bounding Surface in Plasticity Theory

YANNIS F. DAFALIAS

Department of Civil Engineering
University of California
Davis, California 95616 U.S.A.

Summary

The concept of the bounding surface in stress space is presented, and a novel unified formulation of the corresponding rate independent constitutive plastic relations is given. The salient feature of the bounding surface approach is that plastic deformation occurs for stress states within the surface at a rate which depends on the proximity between the current stress point and an "image" stress point on the bounding surface defined by a proper mapping rule. The material exhibits a memory of past loading history by means of plastic internal variables, and a projected "foresight" by means of the distance between actual and "image" stress points. Particular models and corresponding applications are presented within the general framework.

Introduction

The inadequacy of the classical yield surface plasticity formulation to describe realistically the material response under cyclic loading conditions necessitated the development of new concepts [13,15,18]. The purpose of this paper is to give a novel unified presentation of the more recent concept of the "Bounding Surface" [1,4,5,9] which possesses certain attractive features of simplicity, and show the application of different versions to the response of various structural materials subjected to random cyclic loading.

The Bounding Surface Formulation of Rate Independent Plasticity

Assuming the cartesian decomposition of the total strain rate into an elastic and a plastic part $\dot{\varepsilon}^p_{ij}$ and defining the state by means of the stress σ_{ij} and a set of plastic internal variables q_n (e.g., plastic work, plastic strains, residual "back" stresses, etc.), simple rate independent constitutive relations can be given following the basic idea presented in [11] by

$$L = (1/K)\, L_{ij}\, \dot{\sigma}_{ij} \tag{1}$$

$$\dot{q}_n = \langle L \rangle r_n, \qquad \dot{\varepsilon}^p_{ij} = \langle L \rangle R_{ij} \tag{2}$$

where $\langle L \rangle = LH(L)$, H being the step function, L is the loading function, K the plastic modulus, L_{ij} the loading direction (or vector) in stress space and r_n, R_{ij} functions of the state. The inclusion of K in L allows for the description of unstable behavior when $K \leq 0$. It can be shown [12] that the essence of plastic irreversibility requires that $L_{ij} = t(\partial f/\partial \sigma_{ij})$ where $f(\sigma_{ij}, q_n) = 0$ is a loading surface and t an integrating factor. One of the most important objectives, especially for cyclic loading, is the determination of the change of K. In the classical approach the determination of K from the consistency condition $\dot{f}=0$ proved to be ineffective for reverse plastic loading behavior, even for elaborate hardening rules [7].

The bounding surface concept offers a more flexible way for the determination of K which can be illustrated as follows. A typical uniaxial stress-plastic strain curve converges with specific "bounds" as shown in Fig. 1a. The slope E^p of the curve (the uniaxial counterpart of K) can be considered a function of the distance δ between the current state A and a corresponding "image" point $\bar{A}$ on the associated bound, such that $E^p = \infty$ for yielding initiation and $E^p = E^p_o$ when the curve converges with the bound and $\delta = 0$, E^p_o being the bound slope. Such a description has the advantage that once the material response has been "captured" at the extremes (initial yield and "bound"

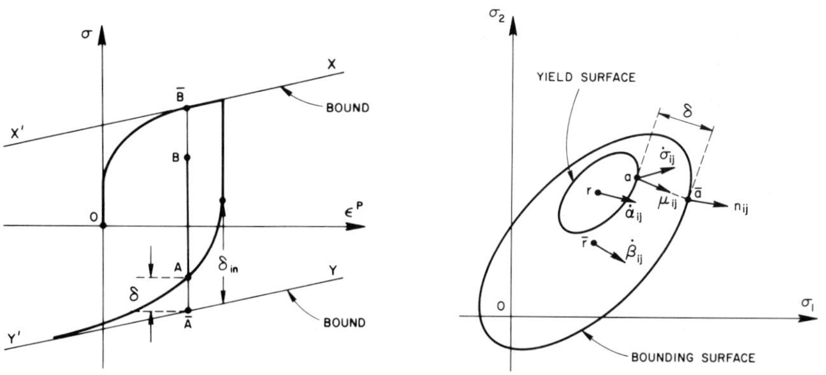

Figure 1a,b. The bounding/yield surface model in uniaxial and multiaxial stress space.

convergence), intermediate state response can easily be obtained. In addition, it is relatively insensitive to the exact definition of the yield point. These notions have recently found an interpretation in terms of dislocations [17].

Generalizing to multiaxial space the points A,B, become a yield(loading surface) and the $\bar{A},\bar{B}$ a "Bounding Surface" enclosing the yield surface, Fig. 1b. One can now formally construct the constitutive relations as follows. The bounding surface is analytically defined by

$$F\,(\bar{\sigma}_{ij},\ q_n) = 0 \tag{3}$$

where a bar over σ_{ij} indicates stresses on F=0. To each σ_{ij} within, a unique "image" point $\bar{\sigma}_{ij}$ on F=0 is defined by a proper, not invertible, mapping rule

$$\bar{\sigma}_{ij} = M_{ij}(\sigma_{k\ell},\ q_n) \tag{4}$$

such that $L_{ij} = (\partial F/\partial\bar{\sigma}_{ij})$ at σ_{ij}. The M_{ij} must satisfy: (a) The integrability condition for the Pfaffian form $(\partial F/\partial\bar{\sigma}_{ij})d\sigma_{ij}$, implying a corresponding loading surface f=0 through σ_{ij}. (b) The identity condition, i.e., M_{ij} must become the identity mapping $\bar{\sigma}_{ij}=\sigma_{ij}$ when σ_{ij} is on F=0. This excludes possible intersection of F=0 and f=0 which would imply that the uniaxial curve crosses the bounds. For any $\dot{\sigma}_{ij}$ such that L>0, a corresponding $\dot{\bar{\sigma}}_{ij}$ occurs through the hardening of F=0 by means of $\dot{q}_n$, and the L is defined now by

$$L = (1/K)(\partial F/\partial\bar{\sigma}_{ij})\dot{\sigma}_{ij} = (1/\bar{K})(\partial F/\partial\bar{\sigma}_{ij})\,\dot{\bar{\sigma}}_{ij} \tag{5}$$

where $\bar{K}$ is the "bounding" plastic modulus (counterpart of E_o^p) associated with $\dot{\bar{\sigma}}_{ij}$. From $\dot{F}$=0 and eqs. (2), (5), one has

$$\bar{K} + (\partial F/\partial q_n)r_n = 0 \tag{6}$$

Extending the uniaxial observations and with $\delta=[tr(\bar{\sigma}_{ij} - \sigma_{ij})^2]^{½}$,

$$K = \hat{K}\,(\bar{K},\ \delta,\ \sigma_{ij},\ q_n) \tag{7}$$

where $\hat{K} > \bar{K}$ for $\delta>0$, $\partial\hat{K}/\partial\delta>0$, $\hat{K} = \bar{K} = K$ for $\delta=0$ and $\hat{K}=\infty$ at yielding initiation. Eq. (7) embodies the salient feature of the bounding surface formulation, i.e., that plastic deformation may occur within F=0 at a rate depending on δ. A classical yield surface formulation can always be obtained in the limit if $\hat{K} \simeq \infty$ for $\delta > 0$ and $\hat{K} \to \bar{K}$ as $\delta \to 0$. In conclusion, one may say that the material exhibits both a memory of past loading history by means of q_n and a projected "foresight" of how far the current state is from a "bounding" state by means of the distance δ.

Particular Models and Applications

The two key features of the bounding surface formulation are the M_{ij} and $\hat{K}$ of eqs. (4) and (7). Their definition specifies different versions as shown in the following. It must be emphasized that each version should be applied to particular classes of material behavior.

(i) The Bounding/Yield Surface Model

The most straightforward approach is to define directly a yield and a bounding surface by $f(\sigma_{ij}-\alpha_{ij}, q_n) = 0$ and $F(\bar{\sigma}_{ij}-\beta_{ij}, q_n) = 0$ [4,5], where α_{ij}, β_{ij} are plastic internal variables accounting for kinematic hardening, Fig. 1b. For congruent surfaces, eq. (4) becomes $\bar{\sigma}_{ij} = c(\sigma_{ij}-\alpha_{ij}) + \beta_{ij}$ with c the congruency ratio. Condition (a) is by definition satisfied and condition (b) imposes restrictions on the coupled hardening of the two surfaces [1,5]. Eq. (7) takes the form

$$K = \left[1 + h(\delta_{in}) \left(\frac{\delta}{\delta_{in}-\delta}\right)\right]\bar{K} \tag{8}$$

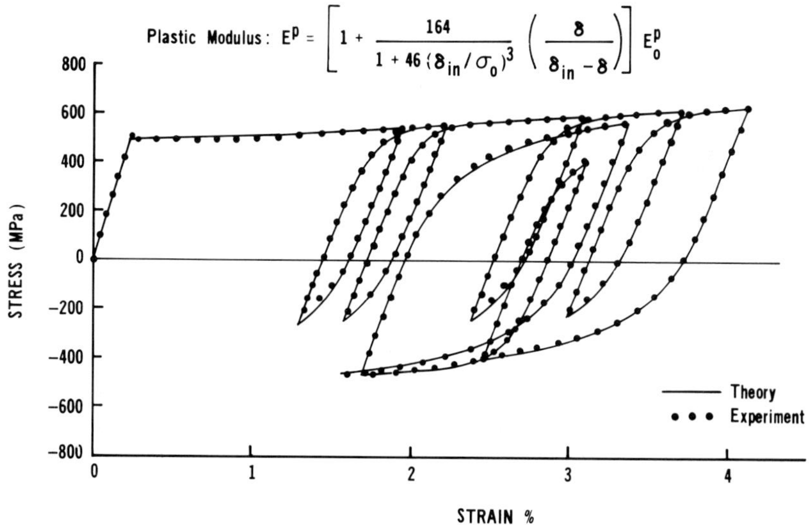

Figure 2. Random cyclic loading of grade 60 steel specimen [5]

where δ_{in} is the value of δ at the initiation of yielding. Application of this model is shown in Fig. 2 where the "plateau" behavior has been treated separately. Similar formulation has been applied to soils [14].

(ii) The "Radial Mapping" Model

This particularly simple model is presented here for the first time and its unique features are illustrated in Fig. 3. One of the q_n is the "back" stress α_{ij} and $\bar{\sigma}_{ij}$ is defined by the intersection of F=0 with the radial line connecting α_{ij} and σ_{ij}. Thus, the M_{ij} takes the form $\bar{\sigma}_{ij} = b(\sigma_{ij} - \alpha_{ij}) + \alpha_{ij}$ with $b \geq 1$ determined from F=0. It can be proved that such mapping satisfies conditions (a) and (b). Eq. (7) takes now the form

$$K = \left[1 + \frac{h}{\langle (r/\delta) - s \rangle}\right] \bar{K} \tag{9}$$

where r is the radial distance between α_{ij} and $\bar{\sigma}_{ij}$ and s is a material parameter controlling the size of the purely elastic domain which is simply determined by eq. (9) from $K=\infty$ when $\delta \geq r/s$. Again, $K=\bar{K}$ when $\delta=0$. The motion of α_{ij} in stress space occurs along the $\alpha_{ij}\sigma_{ij}$ radial line and is controlled by a modulus K_α which for constant $\bar{K}$ must satisfy the condition

$$1 \leq \frac{K_\alpha}{\bar{K}} \leq 1 + \frac{rh}{\langle r - s\delta \rangle} \tag{10}$$

in order to guarantee that K decreases monotonically during radial plastic loading as r and δ change. The bounding surface follows a properly defined kinematic hardening rule. A very interesting feature which has been shown experimentally [16], is that the stress point moves always "ahead" and outside the elastic range crossing it at σ^e_{ij} where $\delta = r/s$ and $K = \infty$, Fig. 3. Subsequently, the stress point crosses different loading surfaces which do not need to be analytically described, but they are simply implied by the corresponding mapping rule. Obviously, no consistency condition is required for the elastic range (or any loading surface), thus the name "elastic nucleus" is preferred from yield surface. Observe that α_{ij} is not necessarily the center of the elastic nucleus but is the center of homeothecy among the different surfaces. In the uniaxial case, a partial unloading/reloading curve shows "lowered" yielding [10] and converges with the continuation of the previous curve if unloading has not occurred. The associated "ratcheting" behavior under superimposed small cyclic amplitutes is always bounded. A simple adaptation of this model to soil mechanics has been developed [3] with s=1 and $\alpha_{ij}=0$, and a corresponding application in the triaxial p-q space is shown in Fig. 4 for undrained cyclic deviatoric loading of clay. The associated soil parameters are explained in [3].

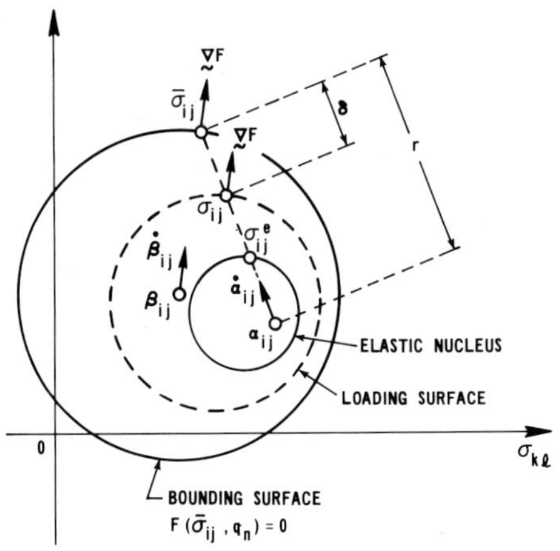

Fig. 3. Schematic illustration of the "radial mapping" model in multiaxial stress space

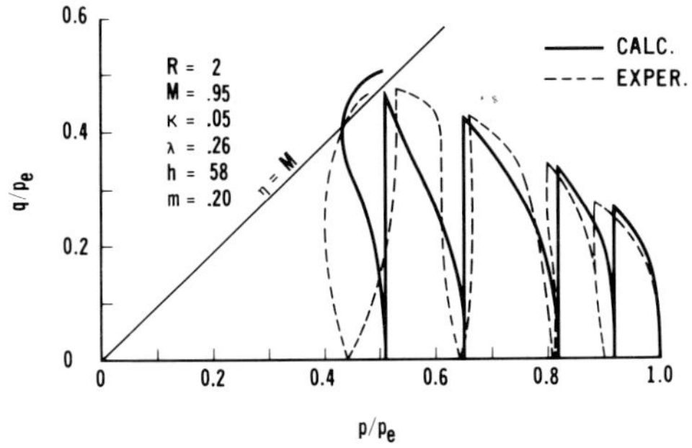

Fig. 4. Application of the "radial mapping" model to undrained cyclic deviatoric loading of clay. Experimental data after [19]

(iii) The Vanishing Elastic Range Model

This is a particular form of the bounding/yield surface model obtained in the limit as the yield surface "shrinks" to zero for materials with no purely elastic range, while the bounding surface remains finite. The determination of the "image" point $\bar{\sigma}_{ij}$ depends now on the state and also on the direction but not the magnitude of $\dot{\sigma}_{ij}$ according to appropriate mapping rules specified in accordance with the kinematic hardening considered for the vanishing yield surface [6,2]. Therefore, the quantities L_{ij}, δ, K and $R_{ij} = L_{ij}$ for normality, depend all on the direction of $\dot{\sigma}_{ij}$. Consequently, the integrability condition has no meaning and plastic loading occurs for any $\dot{\sigma}_{ij}$ with the moduli relating stress and strain rates being homogeneous of order zero in $\dot{\sigma}_{ij}$. This particular form of rate independent plastic irreversibility can be called "Hypoplasticity" and is different from theories with yield surfaces which exhibit "corners" and a more restricted dependence on the $\dot{\sigma}_{ij}$ direction. The concept is illustrated in Fig. 5 for Ziegler's kinematic hardening and an application is shown in Fig. 6 for uniaxial cyclic loading of artificial graphite.

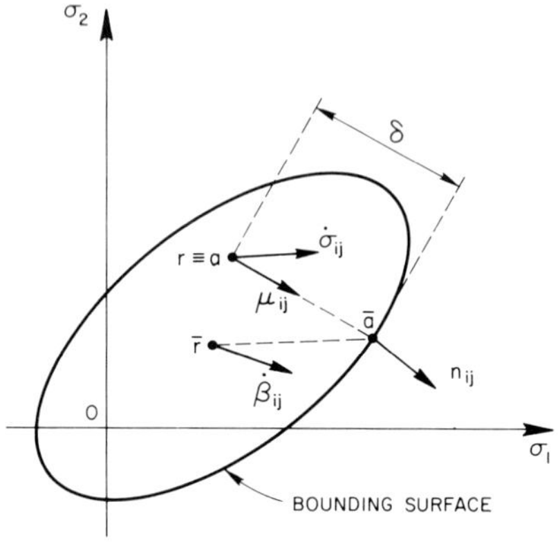

Fig. 5. Illustration of the vanishing elastic range model for Ziegler's kinematic hardening rule.

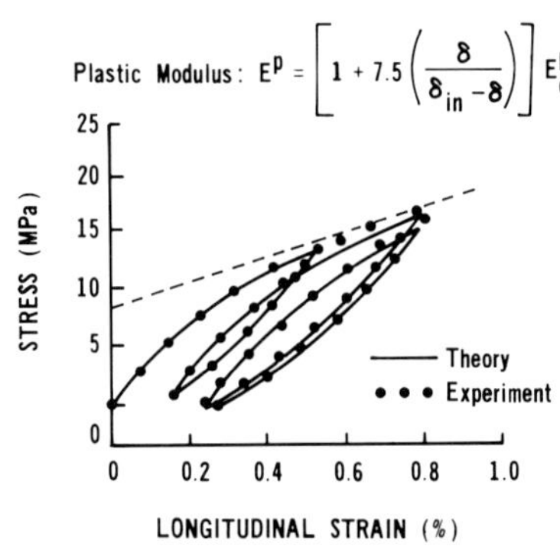

Fig. 6. Cyclic loading of cross-grain-EGCR-type AGOT speciment using the vanishing elastic range model. Experimental data after [8]

References

1 Dafalias, Y.F.: On cyclic and anisotropic plasticity. Ph.D. thesis, University of California, Berkeley (1975).

2 Dafalias, Y.F.: A model for soil behavior under monotonic and cyclic loading conditions. 5th SMIRT, Berlin, K 1/8 (1979).

3 Dafalias, Y.F.; Herrmann, L.R.: A bounding surface soil plasticity model. Int. Symp. Soils under Cyclic Transient Loading, Swansea (1980) 335-345.

4 Dafalias, Y.F.; Popov, E.P.: A model of nonlinearly hardening materials for complex loading. Proc. 7th U.S. National Congress of App. Mech. (1974) 149 and Acta Mechanica 21 (1975) 173-192.

5 Dafalias, Y.F.; Popov, E.P.: Plastic internal variables formalism of cyclic plasticity. J. App. Mech. 98 (1976) 645-650.

6 Dafalias, Y.F.; Popov, E.P.: Cyclic loading for materials with a vanishing elastic region. Nucl. Eng. Design 41 (1977) 293-302.

7 Eisenberg, M.A.; Phillips, A.: On nonlinear kinematic hardening. Acta Mechanica 5 (1968) 1-13.

8 Greenstreet, W.L.; Smith, J.E.; Yahr, G.T.; Valachovic, R.S.: The mechanical behavior of artificial graphites as portrayed by uniaxial tests. Carbon 8 (1970) 649-663.

9 Krieg, R.D.: A practical two surface plasticity theory. J. App. Mech. 42 (1975) 641-646.

10 Lubahn, J.D.: Bauschinger effect in creep and tensile tests on copper. Journal of Metals 7 (1955) 1031-1033.

11 Lubliner, J.: A simple theory of plasticity. Int. J. Solids Struct. 10 (1974) 313-319.

12 Lubliner, J.: On loading, yield and quasi-yield hypersurfaces in plasticity theory. Int. J. Solids Struct. 11 (1976) 1011-1016.

13 Mroz, Z.: On the description of anisotropic work hardening. J. Mech. Phys. Solids 15 (1967) 163-175.

14 Mroz, Z.; Norris, V.A.; Zienkiewicz, O.C.: Application of an anisotropic hardening model in the analysis of elastoplastic deformation of soils. Geotechnique 29 (1979) 1-34.

15 Phillips, A.; Sierakowski, R.L.: On the concept of the yield surface. Acta Mechanica 1 (1965) 30-35.

16 Phillips, A.; Tang, J.L.: The effect of loading path on the yield surface at elevated temperatures. Int. J. Solids Struct. 8 (1972) 463-474.

17 Popov, E.P.; Ortiz, M.: Macroscopic and microscopic cyclic metal plasticity. Proc. 3rd ASCE/EMD Specialty Conference, Austin (1979).

18 Valanis, K.C.: A theory of viscoplasticity without a yield surface, Part 1: general theory, Part 2: application to mechanical behavior of metals. Arch. of Mech. 23 (1971) 517-551.

19 Wroth, C.P.; Loudon, P.A.: The correlation of strains with a family of triaxial tests on overconsolidated samples of Kaolin, Proc. Geotech. Conf. Oslo 1 (1967) 159-163.

Deflexions of Elastic-Plastic Beams at Finite Spread of Plastic Zones

Stanisław DOROSZ, Antoni SAWCZUK

Institute of Fundamental Engineering Research,
Swiętokrzyska 21, 00-049 Warsaw, Poland

Summary

Influence of plastic zones on deflexions of elastic plastic beams is studied. Methods for displacement evaluation and bounding are proposed employing the Betti reciprocity theorem when distorsion field due to plastic deformation appears in finite zones. Comparisons are made with experiments under monotonic as well as variable loading.

1. Introduction

The elastic-plastic analysis and design for metal structures are nowadays admitted in the design codes for beams and stiffened frames [1, 2]. The collapse load evaluation constitutes already a standard procedure [3-5]. Methods allowing to establish or to estimate deflexions which occur prior to collapse or in the elastic-plastic range under service loads as well as at shakedown are developed to a lesser degree.

Available methods of deflexion estimation for elastic-plastic beams and frames concern small displacements. They can be arranged in two groups. In the first one elastic-plastic deformations under monotonically increasing loading are traced as the load increases and plastic zones develop. A structure of variable stiffness is thus analysed until a collapse mechanism forms. The method is straightforward but cumbersome and only solutions regarding simple cases of loading and boundary conditions are available [6, 7].

The other approach studies deflexions at a specific point of the structure transforming into a mechanism of collapse. The beam or frame is considered to be of constant stiffness except at the hinges where no bending stiffness occurs once the yield moment M_o is reached. In this way the finite spread of plastic zones is disregarded and therefore the displacements at collapse are underestimated [3-5]. The method cannot be employed when requirements on displacements are imposed under service loads.

Deflexions of elastic plastic structures under repeated or variable loads are as a rule estimated under the assumption of ideal plastic hinges thus no account is taken of the presence of finite zones of plastic deformations [8].

The paper concerns the influence of plastic zones on deflexions of elastic-plastic beams and frames. Both loads increasing in proportion to a single parameter and loads varying arbitrarily between prescribed limits are considered. Comparisons are made with experiments [9, 10] and with some other methods [11, 12].

2. Problem formulation

We consider structures made of an ideally elastic, perfectly plastic material of the yield stress σ_o. The actual stress σ_{ij}, strain ε_{ij} and displacement u_i at any instant of time are decomposed as follows

$$\sigma_{ij} = \sigma_{ij}^E + \sigma_{ij}^R , \qquad \varepsilon_{ij} = \varepsilon_{ij}^E + \varepsilon_{ij}^R + \varepsilon_{ij}^P , \qquad u_i = u_i^E + u_i^R \tag{1}$$

where σ_{ij}^E, ε_{ij}^E, u_i^E specify the linearly elastic response to the given loads, while σ_{ij}^R, ε_{ij}^R, u_i^R denote the residual stress, strain and displacement, respectively, as caused by an incompatible plastic strain ε_{ij}^P. Therefore

$$\varepsilon_{ij}^{E} = A_{ijkl} \ \sigma_{kl}^{E} , \qquad \varepsilon_{ij}^{R} = A_{ijkl} \ \sigma_{kl}^{R} \tag{2}$$

where A_{ijkl} is the elastic moduli tensor and

$$\varepsilon_{ij}^{E} = u_{(i,j)}^{E} , \qquad \varepsilon_{ij}^{R} + \varepsilon_{ij}^{P} = u_{(i,j)}^{R} \tag{3}$$

Moreover, the following equilibrium requirements apply within the solid and on the surface where tractions P_i are prescribed

$$\sigma_{ij,i}^{E} = 0 , \quad \sigma_{ij}^{E} \ n_j = P_i , \quad \sigma_{ij,i}^{R} = 0 , \quad \sigma_{ij}^{R} \ n_j = 0 \tag{4}$$

Let in an elastic-plastic structure an incompatible plastic strain ε_{ij}^{P} develops under actually acting loads P_i . This incompatible strain field constitutes a distorsion in an elastic solid. To evaluate the elastic-plastic displacements u_i at a point x_i^o we consider a fictitious elastic structure of the same geometry and geometrical constraints. The structure is loaded at x_i^o by unit concentrated force acting in the direction of the displacement looked for. If the stress field in the fictitious elastic solid is $\hat{\sigma}_{ij}$ under the unit load applied at x_i^o then the Betti reciprocity theorem [13] , eventually leads to the following expression for the displacement in the point considered

$$u_i(x_j^o) = \int_V \hat{\sigma}_{ij} \ \varepsilon_{ij}^{E} \ dV + \int_V \hat{\sigma}_{ij} \ \varepsilon_{ij}^{P} \ dV \tag{5}$$

This relation is principal for the developed method and will be used in evaluating displacements of beams when plastic zones are not restricted to the hinges only.

3. Deflexions of beams

Let us consider an elastic-plastic beam of the moment of inertia J, the bending moment $M = M^E + M^R$ where M^E stands for a linearly elastic response and M^R denotes the residual moment due to the plastic strain ε^P. The plastic strains are separated into two contributions. The first one $\bar{\varepsilon}^P(x,z)$ denotes plastic strains in zones of finite extent. This is the distorsion field in partially plastic cross-sections. The second part of the distorsion is concentrated in plastic hinges , say, where a plastic rotation θ_i^P takes eventually place.

Substituting into (5) the respective expressions concerning stresses and strains in beams the following expression is obtained for the deflexion at a loaded point

$$u = \int_L \frac{\hat{M}(x)\, M^E(x)}{EJ}\, dL + \int_L \int_{-H}^{H} \frac{\hat{M}(x)\, z\, B}{J}\, \bar{\varepsilon}^P(x,z)\, dz\, dL + \sum \hat{M}_j \theta_j^P \qquad (6)$$

where $2H$ is the beam depth, B stands for its width and L denotes the respective integration length. Moreover is the elastic moment due to a unit concentrated force acting on a fictitious elastic beam at the point x_0 where the deflexion is looked for according to (6). If the extent of plastic zone is restricted to a plastic hinge the second integral on the right hand side of (6) vanishes.

To illustrate the procedure of accounting for the finite spread of plastic zones let us consider an i-th element as shown in Fig. 1a where M and $\hat{M}$ are marked. The beam is point loaded so that the bending moment distribution is linear. In Fig. 1b the stresses and strains entering (1) are visualized.

For a rectangular cross-section in the partially plastic zone

$$M = \sigma_o B H^2 (3 - \zeta^2)/3 \tag{7}$$

where the boundary $\zeta(x)$ of the elastic kernel is

$$\zeta^2(x) = 3(1 - m_i + m_i x / l_i) , \tag{8}$$

$m_i = M_i / M_o$ and l_i is specified in Fig. 1. The plastic distorsion $\bar{\varepsilon}^P$ is

$$\bar{\varepsilon}^P = \sigma_o (2 - \zeta - H)/\zeta E H \tag{9}$$

and therefore the influence of the i-th plastic region on the deflexion at i is found to be

$$\int_{L_i}\int_{-H}^{H} \hat{M} \frac{\bar{\varepsilon}^P z B}{J} dz\, dL = \frac{\sigma_o \hat{M}_i l_i^2}{E H \hat{l}_i} \Big[\frac{1}{4} m_i (1 - 3 \frac{\hat{l}_i}{l_i}) + \tag{10}$$

$$+ \frac{1}{m_i}(\frac{\hat{l}_i}{l_i} - 1) + \frac{20}{27}\frac{1}{m_i} - \frac{2\sqrt{3}}{3}\frac{\hat{l}_i}{l_i m_i}(1-m_i)^2 - \frac{4\sqrt{3}}{9}\frac{1}{m_i^2} \cdot$$

$$\cdot (1-m_i)^{3/2} \Big]$$

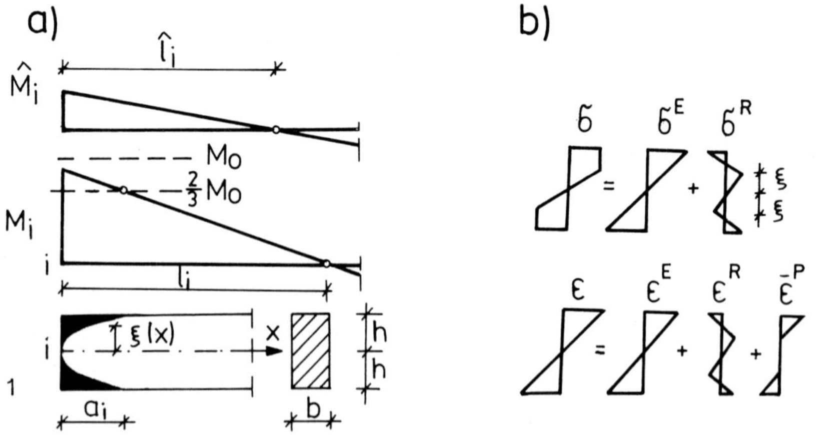

Fig. 1

When $l_i = \hat{l}_i = L$ the deflexion at the end of a cantilever beam is obtained and the result coincide with that following from integration of the flexure equation for a partially plastic beam [3].

4. Estimation of deflexions

Eq. (6) allows to devise a method yielding estimates to deflexions. It consists in bounding the second integral of the right hand, namely for an element with a plastic zone

$$\int_L \int_{-H}^{H} \frac{\hat{M} z B}{J} \bar{\varepsilon}^P dz\, dL \leqslant \max \hat{M} \int_L \int_{-H}^{H} \frac{z B}{J} \bar{\varepsilon}^P dz\, dL = \max \hat{M}\ \theta^* \tag{11}$$

where θ^* denotes the additional plastic rotation due to a finite spread of the plastic zone

$$\theta^* = \int_L \int_{-H}^{H} \frac{z B}{J} \bar{\varepsilon}^P dz\, dL \tag{12}$$

The displacement is therefore estimated as follows

$$u \leqslant u^E + \sum \hat{M}_j \left(\theta_j^P + \theta_j^* \right) \tag{13}$$

where θ_j denotes the rotation at the hinges associated with the actual mechanism of motion.

This justifies the procedure developed in [11]. The beam is there divided into regions $0 \leqslant x \leqslant x_o$, purely elastic, and $x_o \leqslant x \leqslant L$ elastic-plastic. Rotations θ due to the elastic-plastic regions are calculated for various distributions of strains, thus for various bending moments and the cross-section geometry. These rotations are added to the hinge rotations ϕ yielding an estimate

$$u \leqslant \sum \hat{M}\phi + \sum \hat{M}\theta + \sum \int_{x_o}^{L} \frac{\hat{M}\,M}{EJ}\,dx \qquad (14)$$

where summation goes over all the plastic hinges, all the partially plastic zones and all the purely elastic regions.

5. Deflexion bounding

The actual displacement (6) can be approximated from above assessing the extremal values of u_i^E and u_i^R. Elastic response is easily obtained. In [12] it was shown that the permanent displacement u_i^R satisfies the following inequality

$$|u_i^R| \leqslant \max_V \left(\frac{\hat{\sigma}_{ij}(x,x_o)}{\varkappa}\right) \int_V \int_0^t \sigma_{ij}\,\dot{\varepsilon}_{ij}^P\,dt\,dV \qquad (15)$$

where $\varkappa$ depends on the yield condition.

The bound (15) is applicable when loads are repeated or variable. It associates the permanent deflexion with the dissipated energy. Therefore in the case of shakedown

$$\int_0^\infty \int_V \sigma_{ij}\,\dot{\varepsilon}_{ij}^P\,dt\,dV \leqslant \frac{s}{s-1} \int_V \frac{1}{2} A_{ijkl}\,\sigma_{ij}^R\,\sigma_{kl}^R\,dV \qquad (16)$$

where $s > 1$ is the safety factor and σ_{ij}^R stands for a time independent residual stress field required by the Melan theorem.

In the case of monotonic loading and for the shape and extent of the plastic zone as well as for the distribution of plastic strains shown in Fig. 1 the energy dissipated becomes

$$\int_V \sigma_o \bar{\varepsilon}^P dV = 2\sigma_o \int_{L_i}\int_{\zeta H}^{H} \frac{\sigma_o B}{\zeta H E}(z-\zeta H)\,dz\,dL = \frac{2\sigma_o^2}{E} H B L_i$$

$$\left[-1+\frac{10}{9}\frac{1}{m_i} - \frac{1}{\sqrt{3}\,m_i}(1-m_i)^{1/2} - \frac{1}{\sqrt{3}\,m_i}(1-m_i)^{3/2}\right] \qquad (17)$$

For the end joint loaded cantilever beam an upper bound to the maximum permanent deflexion under the monotonic loading is

$$u^R \leq \frac{\max \hat{M}(x,x_o)}{W\,\sigma_o} \int_V \sigma_o \bar{\varepsilon}^P dV =$$

$$= 3\,\frac{\sigma_o L^2}{EH}\left[-1+\frac{10}{9}\frac{1}{m} - \frac{1}{\sqrt{3}\,m}(1-m)^{1/2} - \frac{1}{\sqrt{3}\,m}(1-m)^{3/2}\right] \qquad (18)$$

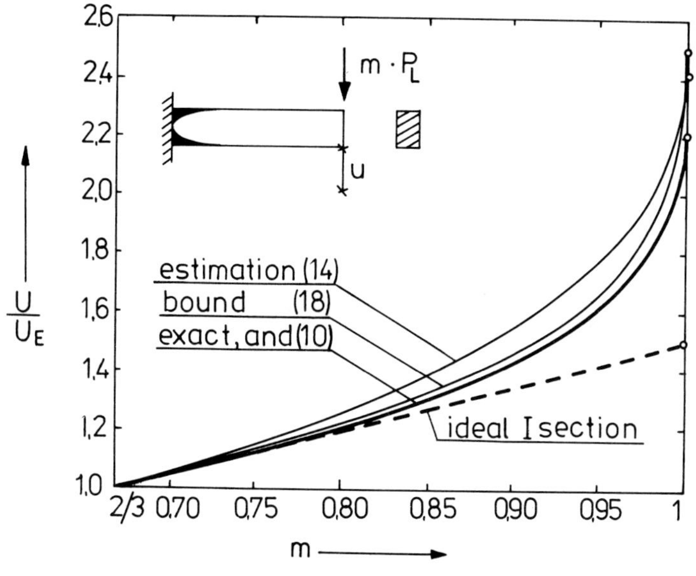

Fig. 2

In Fig. 2 a comparison is made of the developed methods (10), (14), (18). It is seen that the ideal hinge approximation yields to an underestimation of deflexions. The proposed bounding techniques by introducing an additional rotation due to the finite extent of plastic zone preserve the standard procedure of displacement evaluation as developed for the ideal hinge method. For typical bending moment distributions the values of θ^* or θ entering (13) and (14) can be computed.

6. Experiments

Models of continuous beams were tested as to their deflexions in presence of plastic zones [9, 10]. In Fig. 3

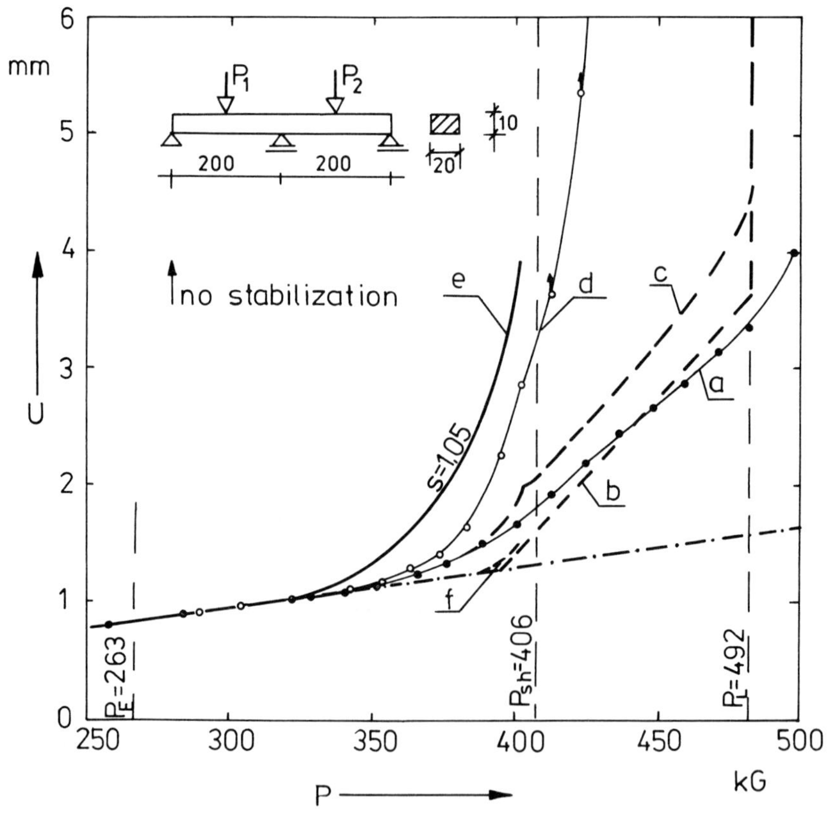

Fig. 3

the results of experiments under uniformly increasing as well as cyclic loadings are compared to the data furnished by the developed methods for deflexion estimation when taking into account the finite spread of plastic zones. The line a concerns experiments under increasing loads. The estimates concerning ideal plastic hinges and finite spreads of plastic zones are given by the curves b and c, respectively. The deflexion under cyclic loading is given by the curve d, whereas e and f indicate a bound obtained in [11] along the lines explained in Sec. 5, when influence of the finite spread of plastic zones was accounted, or neglected, respectively.

Bibliography

1. Recommendations pour le calcul en plasticité des constructions en acier, Constr.metal. No 4, 1975
2. PN-76/B-03200 Steel Structures. Analysis and Design /in Polish/, Wydawnictwo Normalizacyjne, Warszawa 1977
3. P.G.Hodge,Jr, Plastic Analysis of Structures, McGraw-Hill, New York 1959
4. J.Heyman, Plastic Design of Frames. Applications, Cambridge University Press, Cambridge 1971
5. Ch.Massonnet, M.Save, Calcul plastique des constructions, Vol.1, Nellisen, Liège 1976
6. A.R.Rzhanitsyn, Analysis of Structures Taking into Account Plastic Properties of Materials /in Russian/, Stroyizdat, Moscow 1954
7. K.A.Reckling, Plastizitätstheorie und ihre Anwendung auf Festigkeitsprobleme, Springer Berlin 1967
8. J.A.König, Shakedown deflections, Teoret.Pril. Mekh. 3, 1972, 65-69
9. S.Dorosz et al., Post-yield deflections of elastic-plastic beams under uniformly increasing loads, Rozpr. Inż. 1980 /in press/
10. S.Dorosz et al., Deflections of elastic-plastic hyperstatic beams under cyclic loading, Rozpr.Inż./submitted/
11. W.Bandyszewski, A.Sawczuk, Deflection estimation for elastic-plastic beams and frames at finite spread of yielding zones, Arch.Inz.Ląd. 1980 /in press/
12. S.Dorosz, Estimation of deflections of elastic-plastic structures at shakedown /in Polish/ IPPT Reports, 65/77 Warsaw 1977
13. W.Nowacki, Theory of Elasticity /in Polish/, PWN Warsaw 1970

Axial Strain in Cyclic Torsion

R. N. DUBEY

Department of Mechanical Engineering
University of Waterloo, Waterloo, Ontario

Abstract

A theory to obtain axial strain and shear angle in a thin tube under torque is developed. A feature of the analysis is that it takes into consideration the motion of typical material line elements as the body deforms from an initial to a final configuration. The axial stretch and shear angle during a certain cycle of loading are both found to depend upon their values at the end of the previous cycle.

Introduction

One of the purposes of this paper is to develop a scheme to calculate axial strain in a thin circular tube under cyclic torsion. We propose to use the constitutive model of the incremental theory of elastic-plastic deformation. The theory uses a smooth yield surface which expands isotropically for a strain-hardening material. Because of the normality of the flow to the yield stress, the theory predicts no axial strain for a cylinder under torque (zero axial load). Thus, the available evidences show that our attempt must fail. Yet, we proceed. We expect to succeed. There are several reasons for this expectation.

The variables used in the constitutive model of the incremental theory are always measured on the instantaneous (current) configuration which, unfortunately, keeps changing during continued deformation. The incremental strains calculated at different instants during a sequence of incremental loading refer to different configurations in the sequence of deformation. The incremental strain at a particular instant is a true measure of change of configuration at that instant only. It seems that the sum of these increments would yield correct strain values for special measures of strain only and even that may be true for special cases only. In general, summation or integration of strain-increments yield values which are hard to interpret. In any event, these values should not, except in special circumstances, be used for the purpose of predicting the final configuration of a body from a given initial configuration and for a given loading history.

We propose to use the following technique to calculate the final strain. Consider a line element, or a set of two line elements. The incremental strain in these elements is calculated at each instant of time. This may be integrated to yield final longitudinal strain in the element or the final shear angle between the two line elements. A complete explanation of this statement follows.

Current Configuration and Stress

Let us assume a homogeneous and plane stress distribution in the tube. We further assume that a plane section normal to the tube axis remains plane and normal throughout the deformation. As a result, a point initially at X_1, X_2 (= $R\theta$) on the tube surface is displaced to

$$\begin{aligned} x_1 &= a_1 \cos\gamma X_1 , \\ x_2 &= a_1 \sin\gamma X_1 + a_2 X_2 , \end{aligned} \tag{1}$$

where a_1 is the stretch of a line element which was initially parallel to the tube axis and γ is the angle between the two directions of this element; a_2 is the ratio of the final and initial radii of the tube.

The current distribution of Cauchy (true) stress is obtained from (1) and the formula

$$\sigma_{ij} = \frac{\rho}{\rho_o} \frac{\partial x_i}{\partial X_k} S_{kj} , \tag{2}$$

where S_{ij} is the nominal stress and ρ, ρ_o are the current and initial densities. We simplify (2) by assuming constancy of volume and obtain

$$\sigma_{ij} = \frac{\partial x_i}{\partial X_k} S_{kj} \ . \tag{3}$$

Suppose the traction boundary condition is such that

$$T_1 = \nu_1 S_{11} = 0 \ ,$$
$$T_2 = \nu_1 S_{12} = S_{12} \ . \tag{4}$$

The hoop stress

$$S_{22} = 0 \tag{5}$$

because of the equilibrium condition and the homogeneous plane stress distribution. The components of the true stress, from (1), (3), (4) and (5), are

$$\sigma_{11} = 0 \ ,$$
$$\sigma_{12} = a_1 \cos\gamma S_{12} = \tau \text{ (say)} \ , \tag{6}$$
$$\sigma_{22} = a_1 \sin\gamma S_{12} = \tau \tan\gamma \ .$$

The value of a_1 and γ is obtained with the help of the equations relating the incremental quantities.

Incremental Deformation

The constitutive equation for an isotropic and incompressible elastic-plastic solid may be expressed in the form

$$de_{ij} = \sigma'_{ij}(d\bar{\sigma}/h\bar{\sigma}) + (d\sigma_{ij} - d\sigma_{kk}\delta_{ij}/3)/2\mu \ , \tag{7}$$

where de_{ij} is strain increment, $d\sigma_{ij}$ is an objective stress increment, σ'_{ij} is the deviatoric component of the true stress, $\bar{\sigma}$ is the effective stress and μ and h are, respectively, the elastic and plastic moduli.

The inverted form of (7) is

$$d\sigma_{ij} = dp\delta_{ij} + 2\mu\{de_{ij} - (2\mu/h + 2\mu)m_{ij}m_{k\ell}de_{k\ell}\} \tag{8}$$

where dp is an unknown hydrostatic pressure, and

$$m_{ij} = \sigma'_{ij}/(\sigma'_{k\ell}\sigma'_{k\ell})^{1/2} . \tag{9}$$

Finally, the objective stress increment and the nominal stress increment, dS_{ij}, are related by

$$d\sigma_{ij} + \frac{\partial u_j}{\partial x_k}\sigma_{ik} - (de_{ik}\sigma_{jk} + de_{jk}\sigma_{ik}) \equiv \frac{\partial x_i}{\partial X_k} dS_{kj} , \tag{10}$$

where u_i is the incremental displacement.

In view of our earlier assumption that a plane normal to the axis remains plane and normal, we have

$$\frac{\partial u_1}{\partial x_2} = 0 ,$$

and (11)

$$\frac{\partial u_2}{\partial x_1} = 2de_{12} .$$

We also have

$$dS_{11} = 0 , \quad dS_{22} = 0 \tag{12}$$

which follows from a consideration of the incremental boundary and equilibrium conditions. On substituting from (1), (6), (11) and (12), equation (10) yields

$$d\sigma_{11} - 2\tau de_{12} \equiv 0 ,$$

$$d\sigma_{12} - \tau de_{11} - \tau\tan\gamma de_{12} \equiv a_1\cos\gamma dS_{12} = d\tau \text{ (say)} \tag{13}$$

$$d\sigma_{22} - \tau\tan\gamma de_{22} \equiv a_1\sin\gamma dS_{12} = \tan\gamma d\tau .$$

Elimination of $d\sigma_{ij}$ from (8) and (13) results in three equations involving three unknowns de_{11}, de_{12} and de_{22}. An approximate solution of these equations, correct up to third order terms, are

$$de_{11} = -\frac{1}{3}\frac{2\mu + h}{2\mu h}\{\tan\gamma - 2\frac{\tau}{2\mu}\}\,d\tau ,$$

$$de_{22} = \frac{2}{3}\frac{2\mu + h}{2\mu h}\{\tan\gamma - \frac{\tau}{2\mu}\}\,d\tau , \tag{14}$$

$$de_{12} = \frac{2\mu + h}{2\mu h}\,d\tau .$$

A note of warning ought to be put in here. We c a n n o t obtain the axial strain by integrating de_{11} in (14). In fact, none of three values obtained by integrating de_{ij} in (14) can be interpreted as strain required to deform the cylindrical tube from the initial to the final configuration. For this purpose, we need to choose a set of material line elements; calculate incremental strain, $d\varepsilon_{ij}$, in these elements. The integrated values of $d\varepsilon_{ij}$ yield information regarding the total change of configuration.

History of Deformation

Let us choose two material line elements such that they are initially orthogonal, one parallel and the other perpendicular to the axis, and of lengths DX and ΔX. In the current configuration, their vector components, respectively, are

$$\begin{aligned} Dx_i &= (a_1 DX) n_i , \\ \Delta x_i &= (a_2 \Delta X) t_i . \end{aligned} \tag{15}$$

The component of the unit vectors $\underset{\sim}{n}$ and $\underset{\sim}{t}$ are

$$\begin{aligned} n_1 &= \cos\gamma , \quad & n_2 &= \sin\gamma ; \\ t_1 &= 0 , \quad & t_2 &= 1 . \end{aligned} \tag{16}$$

From calculus, we have

$$Dx_i \Delta x_i = Dx\Delta x\cos(\frac{\pi}{2} - \gamma) . \tag{17}$$

The above equation yields the following relationship between incremental shear angle and strain:

$$2de_{ij}n_i t_j = (de_{\underset{\sim}{n}} + de_{\underset{\sim}{t}})\sin\gamma + \cos\gamma d\gamma \tag{18}$$

where

$$de_{\underset{\sim}{n}} = de_{ij}n_i n_j$$
$$de_{\underset{\sim}{t}} = de_{ij}t_i t_j \tag{19}$$

are, respectively, the incremental strain in the elements DX and ΔX in the current configuration.

Substituting (14) in equation (18) and using (16), we obtain

$$\frac{2\mu + h}{2\mu h}\, d\tau = d\gamma/(1 + \cos^2\gamma) \tag{20}$$

which is correct up to $(\tau/2\mu)^2$ terms. The integration of (20) yields

$$\tan\gamma = \sqrt{2}\, \tan(\sqrt{2}x + C_1) \tag{21}$$

where we have used the notation

$$x \equiv \int \frac{2\mu + h}{2\mu h}\, d\tau \tag{22}$$

and C_1 is a constant.

At the beginning of the first cycle of loading, $x = 0$ and $\gamma = 0$ and hence $C_1 = 0$. In this case, (21) simplifies to

$$\tan\gamma = \sqrt{2}\, \tan(\sqrt{2}x) \ . \tag{23}$$

During subsequent cycles, at the beginning $\gamma = \gamma_n$ (say) when $x = 0$. The value $C_1 = C_1^{(n)}$ is then obtained from

$$\tan\gamma_n = \sqrt{2}\, \tan C_1^{(n)} \ . \tag{24}$$

It follows that, during this cycle,

$$\tan\gamma = \sqrt{2}\, \tan\{\sqrt{2}x + \arctan(\frac{1}{\sqrt{2}} \tan\gamma_n)\} \ . \tag{25}$$

Equation (25) can be used to calculate γ in the nth cycle of torsional loading of a tubular specimen.

In order to calculate axial strain, we substitute from (14) and (16) into (19), and use the definition of Eulerian strain increment to obtain

$$\frac{da_1}{a_1} = de_{\underset{\sim}{n}} = \frac{2\mu + h}{6\mu h}\{\tan\gamma(2 + 3\cos^2\gamma)\}d\tau , \tag{26}$$

up to $(\tau/2\mu)^2$ terms. From (20) and (26), we have

$$\frac{da_1}{a_1} = d\gamma\frac{\tan\gamma(2 + 3\cos^2\gamma)}{3(1 + \cos^2\gamma)} . \tag{27}$$

The above equation can be integrated to yield

$$a_1^2 = C_2\sec^2\gamma(1 + \frac{1}{2}\tan^2\gamma)^{-1/3} , \tag{28}$$

where C_2 is a constant. At the beginning of the first cycle of loading, $\gamma = 0$, $a_1 = 1$, and therefore $C_2 = 1$. In this case,

$$a_1^2 = \sec^2\gamma(1 + \frac{1}{2}\tan^2\gamma)^{-1/3} . \tag{29}$$

At the beginning of the nth cycle, let

$$x = 0 , \quad a_1 = a_1^{(n)} , \quad \gamma = \gamma_n .$$

The constant C_2 is then given by

$$C_2^{(n)} = (a_1^{(n)}\cos\gamma_n)^2(1 + \frac{1}{2}\tan^2\gamma_n)^{1/3} \tag{30}$$

and

$$a_1^2 = C_2^{(n)}\sec^2\gamma(1 + \frac{1}{2}\tan^2\gamma)^{-1/3} . \tag{31}$$

Equation (31) may be used to calculate axial strain in the nth cycle of torsional loading of a tubular specimen.

Acknowledgement

The research was partly supported by NSERC Grant No. A5164.

The work is based on a Ph.D. Thesis of a former student, Dr. H. P. Srivastava. The reference to the thesis is gratefully acknowledged.

Effect of Physical Nonlinearities on Optimal Design of Structures

A. GAJEWSKI

Institute of Physics
Technical University
of Cracow, Poland

Summary
The subject of the paper is the minimal weight design of structural elements made of material characterized by a nonlinear ralations between stress and strain tensors. In a wide class of problems the form of the physical law either has no influence on the optimal shape or affects it only by a constant factor.

Formulation of the problem

The problem of the structural optimal design can be understood as a certain operation A which attributes to various elements of a set X elements of another set Y:

$$A : X \longrightarrow Y \tag{1}$$

The elements of set X are relations determined on the fields: 1. mechanical-state variables $(\underset{\sim}{\sigma}, \underset{\sim}{\varepsilon}, \underset{\sim}{u})$, 2. optimization parameters (control variables ϕ_i), 3. external and internal loadings. We assume that the space X is the Cartesian product of five subsets of the ralations:

$$X = X_1 x\ X_2 x\ X_3 x\ X_4 x\ X_5 . \tag{2}$$

X_1 is the subset of the physical laws. It consists of the equations connecting the mechanical-state variables. In the case of nonhomogeneous material these equations are also dependent on the point $P(x,y,z)$:

$$\underset{\sim}{\varepsilon} = \hat{F}(\underset{\sim}{\sigma}, P), \tag{3}$$

where $\hat{F}$ is a certain operator. X_2, X_3, X_4 and X_5 are the subsets of boundary conditions, body forces distributions, constraints and cost functions respectively. The set Y consists of certain functions describing the optimal shapes of structures. These functions can be represented in three-dimensional space (e.g. in spherical coordinates) by:

$$r = r(\varphi, \vartheta). \quad (4)$$

The aim of this paper is to examine the particular case, namely the dependence of the set Y (optimal shapes) on the subset X_1 (physical laws), provided that the elements of other subsets are constant. Further we shall distinguish three particular cases. 1. The case in which the same element of the set Y corresponds to each of all elements of the subset X_1. Here, the optimal shape of the structural element is independent of the physical law:

$$r = r_o(\varphi, \vartheta). \quad (5)$$

2.The case in which the dependence of the optimal shape on physical laws is determined by the set of a certain number of constants C_i:

$$r = r_o(\varphi, \vartheta, C_i, X_1). \quad (6)$$

We call this kind of dependence inessential. 3.The case in which for various elements of the subset X_1 we obtain also various functions $r(\varphi, \vartheta)$. We call this kind of dependence essential.

Physical laws

In this paper we restrict our considerations to the case of the similarity of stress-strain deviators. These relations can describe non-linear elastic, elastic-plastic as well as rheological behaviour of materials. The scalar factor in stress-strain (or strain rate) relations should be obtained from the nonlinear physical equations:

$$\varepsilon_i/\varepsilon_o = f(\sigma_i/\sigma_o)\, \sigma_i/\sigma_o = \mathfrak{F}(\sigma_i/\sigma_o)$$

or (7)

$$\sigma_i/\sigma_o = g(\varepsilon_i/\varepsilon_o)\, \varepsilon_i/\varepsilon_o = \mathfrak{G}(\varepsilon_i/\varepsilon_o),$$

where:

$$\varepsilon_i = \left(\tfrac{2}{3}\, e_{ij}\, e_{ij}\right)^{1/2}, \qquad \sigma_i = \left(\tfrac{3}{2}\, s_{ij}\, s_{ij}\right)^{1/2}$$

The nonhomogeneity parameters ε_o and σ_o are prescribed functions of (x,y,z).

Formulation of the optimization problem and the method of solution

The weight of the structural element is assumed as the cost function. The constraints can be divided into four kinds: 1. the bounds of the stress or strain intensity, 2. the bound of the integral:

$$I = \int_V \psi(\underset{\sim}{\sigma}, \underset{\sim}{\varepsilon}, \underset{\sim}{u}, p_i, \phi_i, x, y, z)\, dV, \tag{8}$$

where ψ is a certain function, 3. constraints in the form of equilibrium and compatibility equations, 4. buckling constraints. As the functionals (8) we assume the quantities characterizing the stiffness (or compliance) of a construction, for example: the total work (power) of the external loadings $\tilde{A}$, the total strain energy $\tilde{U}$, the total potential energy $\tilde{V}=\tilde{U}-\tilde{A}$, the total complementary energy (power) $\tilde{\Lambda}$, the displacement at a given point, the maximal displacement etc.

The method of the solution is based on the classical variational approach applied to the dual formulation of the problem.

Selected simple examples

1. Optimal design of a compressed column (without buckling)

Provided that the function ψ depends only on the variable σ and the weight density is constant we obtain the solution: $\sigma' = 0$, (σ = const.), which is independent of the form of function $\psi(\sigma)$. Consequently, assuming that ε_o and σ_o are constants (homogeneous material), the optimal shape of the column of constant weight and of minimal free-end displacement is independent of the physical law (7) (details are presented in [1] and [2]).

2. Optimal design of cantilever beam loaded by the external loadings and its own weight.

In this case, presented in the paper [3], the particular solution can be expressed in the form which depends on physical law only through constant parameter.

3. Optimal design of the compressed rods with buckling constraints.

In these cases considered in details in several papers (e.g. [4], [5], [6]), the optimal shapes always depend on physical laws and this dependence is essential.

Conclusions

It is demonstrated in [1] - [3] that there exists a broad class of problems in which the optimal shape of structural element is independent of physical law and is the same as that obtained in linearly-elastic case. One can conclude that this independence appears in particular cases of optimal design of beams and plates made of homogeneous material and not loaded by its own weight if either the total potential energy or the total complementary energy are taken as the cost functions. However, the presentation of one general criterion which allows to judge wether the optimal shape depends on physical law, or not, is very difficult.

Further details can be found in [1].

Referencies

[1]. Gajewski, A.: Optimal structural design in the case of physically non-linear materials in polish , Technical University of Cracow, Scientific Report No.5, Cracow, 1975.

[2]. Gajewski, A.: Optimal structural design of the rotating rod with physical non-linearity of material in polish , Theor.Appl.Mech., 14, 1976, 262-271.

[3]. Albińska, M.; Gajewski, A.: Optimal design of the cantilever beam loaded by external forces and its own weight in creeping conditions in polish , Theor. Appl.Mech., 16, 1978, 495-506.

[4]. Krzyś, W.: Optimale Formen gedrückter dünnwandiger Stützen im elastisch-plastischen Bereich, Wiss.Z.TU Dresden, Internationale Stahlbautagung, 17, 1968 , 407-410.

[5]. Gajewski, A.; Życzkowski, M.: An optimal forming of a bar compressed with subtangential force in elastic-plastic range, Arch.of Mechanics, 23, 1971, 147 - 165.

[6]. Rasmussen, S.H.: On the optimal shape of an elastic-plastic column, J.Struct.Mech., 4, 1976, 147-152.

A Comparison of the Plastic Buckling Behaviour of 2:1 Ellipsoidal and 2:1 Torispherical Shells Subjected to Internal Pressure

G. D. GALLETLY

Department of Mechanical Engineering
University of Liverpool
Liverpool, U.K.

Summary

The elastic-plastic buckling pressures of three (the 'optimum' and two others) 2:1 torispherical shells subjected to internal pressure are compared with the corresponding p_{cr} for the 2:1 ellipsoidal shell. This is done for 207 $N/mm^2 \leq \sigma_{yp} \leq 620$ N/mm^2 and over the diameter-thickness range $750 < {}^D/t < 1250$. The result of the comparison is that the 2:1 ellipsoidal shape appears to have about 25% more buckling resistance than the 2:1 torispherical shape. Whether one can rely on this 25% margin in practice depends on the accuracy of the shape verification procedure. Some examples are discussed to illustrate the problem.

1. Introduction

A review of the buckling and collapse of internally-pressurized perfect dished ends was given recently by Galletly [1] but, in the present work, only shapes with a major/minor axis ratio of 2:1 will be considered. One of the aims was to see if the elastic-plastic internal buckling pressure of a 2:1 ellipsoidal shell was higher than that of 2:1 torispherical shells. While there does not seem to be much work reported in the literature on their respective buckling pressures, Bushnell [2] does give the results of one elastic-plastic calculation.

Even though the perfect 2:1 ellipsoidal shape may be stronger than the 2:1 torispherical one on paper, one has to take into account the fact that welded dished ends in practice are imperfect. The American (ASME, Section VIII, UPV) Code requirement on the shape of formed heads permits the actual shape to deviate from the specified one by up to 1.25% of the inside diameter of the head skirt. The deviations have to be measured perpendicular to the specified shape and should be outside it. Because of these tolerances, it is possible for a 2:1

ellipsoidal head to be delivered on site as 2:1 torispherical; if this should actually occur in practice, it means that the factor of safety will be lower than is thought.

Because of the foregoing, Esztergar [3] suggested that shape measurements in practice should limit the permitted discontinuity in meridional curvature to a low value. The reasoning behind this was that an ellipsoidal shell is stronger than a torispherical one because its curvature varies continuously. This suggestion, as it relates to buckling, will be examined briefly later.

2. Characteristics of the Shells Analysed

The three 2:1 torispherical shells considered in the buckling analysis had the following geometric ratios:

Shell No.	$^r/_D$	$^{R_s}/_D$	$\frac{R_s - r}{D}$	$^r/_{R_s}$
I	0.1458	0.800	0.6542	0.1825
II	0.1726	0.904	0.7314	0.1914
III	0.1875	1.000	0.8125	0.1875

In the above, r is the radius of the torus, R_s the radius of the spherical cap and D the diameter of the cylinder which is being enclosed. Shell II is the so-called 'optimum' torisphere.

The $^D/t$-ratios considered were 750, 1000 and 1250 and the stress-strain curves of the perfect steel shells were assumed to be elastic-perfectly plastic. The yield points of the steel varied from 207 N/mm^2 to 620 N/mm^2.

For the small study on the effect of a discontinuity in the meridional radius of curvature in 2:1 shapes, only $^D/t = 750$ was investigated and σ_{yp} was taken as 414 N/mm^2.

3. Internal Buckling Pressures

The computed p_{cr}'s (obtained from the shell buckling computer program BOSOR 5 [4]) for the 2:1 ellipsoidal and torispherical shells are tabulated below:

$^D/_t$	Shell Type	σ_y (N/mm^2)		
		207	414	620
750	2:1 Tori. I 2:1 Tori. II 2:1 Tori. III	No Buckling 0.588 0.557	0.789 0.858 0.858	1.148 1.231 1.224
	2:1 Ellip.	No Buckling	1.089	1.603
1000	2:1 Tori. I 2:1 Tori. II 2:1 Tori. III	0.298 0.329 0.327	0.533 0.562 0.548	0.824 0.886 0.872
	2:1 Ellip.	0.403	0.772	1.169
1250	2:1 Tori. I 2:1 Tori. II 2:1 Tori. III	0.223 0.230 0.222	0.408 0.433 0.422	0.639 0.676 0.667
	2:1 Ellip.	0.296	0.586	0.927

TABLE I Elastic-Plastic Internal Buckling Pressures (N/mm^2)

From Table 1, it may be seen that the p_{cr}'s of the three torispherical shells are not very dissimilar and that those of Torisphere II (the 'optimum') are marginally the highest. Also, the p_{cr}'s of the 2:1 ellipsoidal shell are higher than those of the optimum 2:1 torispherical shell by amounts which vary from 23% to 37%.

4. Shapes of 2:1 Ellipsoidal and Torispherical Shells

To indicate the problems which might arise in practice in trying to determine whether a shell is a 2:1 ellipsoidal one or not, the following tabulation lists the radial deviations between torispheres I, II and III and a 2:1 ellipsoidal shell, expressed as a percentage of the cylinder diameter. As may be seen, shell II is within the tolerance allowed by the ASME Code and the other two shells very nearly satisfy the requirements.

Angle in Torus, from the Horizontal (Deg.)	Approx. Radial Deviations (%)		
	Shell I	Shell II	Shell III
30	0.19	0.62	0.90
40	0.18	0.83	1.26
50	0.05	0.90	1.43
60	-0.28	0.73	1.38

5. The Effect of a Discontinuous Meridional Radius of Curvature

It was suggested in [3] that the discontinuity in meridional radius of curvature in a 2:1 torisphere might be the cause of its strength being lower than that of the 2:1 ellipsoidal shell. A very limited examination of this suggestion was made herein by taking an optimum 2:1 torispherical shell, removing the portion of the torus between its mid-point and the torus/sphere junction, replacing it by an "insert" shell and then putting back the spherical cap. The insert was chosen to have the same slope and curvature as the torus at its lower end and the same as the spherical cap at its upper end. The shape was found numerically using Lagrangian interpolation coefficients and, while the shape of the resulting insert appeared to be smooth, the curvature in the insert did not vary monotonically. The p_{cr} of the assembly turned out to be very similar to that of the 2:1 optimum torisphere.

One of the other inserts studied was part of a 2:1 ellipsoidal shell. Its meridional radius of curvature matched that of the torus at the lower end but was 2.5 times that of the spherical cap at the upper end (i.e. there was a discontinuity in the curvature there). The p_{cr} for this case was only about 5% lower than that of the 2:1 ellipsoidal shell.

Thus, the results of the above limited study on curvature are a little ambivalent. It is hoped to study the matter further in the future.

6. References

1. Galletly, G.D.: Buckling and collapse of thin internally-pressurized dished ends. Proc. ICE, 67, Part 2, (Sept. 1979), 607-626.
2. Bushnell, D.: Non-symmetric buckling of internally pressurized ellipsoidal and torispherical elastic-plastic pressure vessel heads. ASME (J. Press. Vess. Tech.), 99, (1977), 54-63.
3. Esztergar, E.P.: Development of design rules for dished pressure vessel heads. WRC Bull. 215 (May 1976), 25-30.
4. Bushnell, D.: BOSOR 5 - program for buckling of elastic-plastic shells of revolution including large deflections and creep. Comput. Struct., 6, (1976), 221-239.

Shakedown of Plates under Moving Patch Loads

Paul Grundy and W A M. Alwis
Department of Civil Engineering,
Monash University.

Summary

Shakedown criteria for plates under moving patch loads are derived using a normal moment yield criterion. The formulation results in the determination of incremental collapse and alternating plasticity limits independent of each other. In developing a numerical method, the plate and the load program are discretised and shakedown criteria are linearised. The incremental collapse limit is computed using a linear programming procedure and knowing the active constraints the feasible collapse mechanism can be determined approximately. A simple inequality is used to obtain the alternating plasticity limit. The upper bound formulation is used to obtain the shakedown limit by trial yield line collapse mechanisms, providing a quicker method of calculation for simple cases.

1. INTRODUCTION

The shakedown limit load of frames and grids can be readily determined by well established techniques. In many practical cases where the shakedown limit is close to the static collapse limit, considerations of the latter overshadow the former, so that the shakedown limit is not an important parameter in design. The same does not hold for plates. Here the range between the elastic limit and static collapse is large, and the shakedown limit is not generally known. For bridge decks and similar plates subjected to moving patch loads, the shakedown limit as a bound on damage assumes an importance equal to the static collapse limit. An historical analysis, following the nonlinear response of the plate is impractical since neither the initial stress state nor the load path is known. Mathematical programming (MP) combined with the theory of Engineering Plasticity (EP) offers a means of finding the shakedown limit.[1]

In this analysis the loading is completely defined by a load program of duration τ, repeated indefinitely. The load vector is of the form $\mathbf{p}(t) = \lambda \mathbf{W}.\mathbf{q}(t)$; $t \in \Omega$, where λ is a scalar load multiplier, **W** is the matrix of independent load vectors, $\mathbf{q}(t)$ is the vector of time functions, and Ω is the set of all t (time) in the domain $0 \leq t < \tau$. The shakedown limit is given by the maximum value of λ for which the structure can shakedown.
To derive shakedown criteria we use as a basis the normal moment criterion for yielding of plates, which is widely used in limit analysis. IC and AP limits can be determined independently with this criterion.

2. SHAKEDOWN CRITERIA FOR PLATES

Melan's theorem of shakedown is interpreted as requiring a state of residual stresses in equilibrium with zero loads, that would enable the plate to behave purely elastically under one application of the load program, if the plate is to shakedown. The only stress resultants involved are plate moments. Membrane stress resultants are neglected - a conservative assumption. The state of moments $\{M\}$ defined for a point on the plate is described by its component moments M_{ij}; $i,j = 1,2$ associated with the Cartesian frame (x_1,x_2,x_3) where the x_3 axis is normal to the plate.

2.1 Residual Stresses.

For residual moments R the equilibrium requirement for residual stresses can be expressed in the familiar form:

$$\frac{\partial^2 R_{11}}{\partial x_1^2} + 2\frac{\partial^2 R_{12}}{\partial x_1 \partial x_2} + \frac{\partial^2 R_{22}}{\partial x_2^2} = 0 \qquad [1]$$

2.2 Yield Criterion

Applying the normal moment criterion, which states that the normal moments are equal to normal moment capacities along a yield line of a plate, the incremental plastic rotation (IC) for a plate section S can be expressed in the form

3.

$$Q_n^- \leqslant M_n^\gamma \leqslant Q_n^+; \quad t_\gamma \in \Omega \qquad [2]$$

where Q^-_n and Q^+_n are moment capacities for negative and positive plastic curvatures of S, and M_n^γ is the normal moment at S under the loading $\mathbf{p}(t_\gamma)$.

Extending the normal moment criterion to postulate that only normal stress variations can cause AP, the AP criterion for S is

$$M_n^\mu - M_n^\nu \leqslant T_n; \quad t_\mu, t_\nu \in \Omega \qquad [3]$$

where T_n is the permissible variation in normal moments at S so that no plastic deformations take place. For elastic-plastic materials T_n is a constant for a section.

3. MATHEMATICAL FORMULATION OF YIELD CRITERION

The criteria [2] and [3] expressed for a plate section S have been mathematically developed[5,13] to arrive at inequalities that refer to a point on the plate.

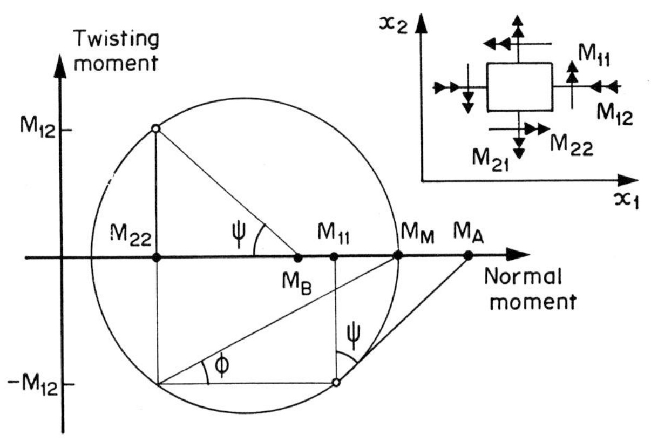

Fig. 1 Mohr Circle of State of Moments {M}

Our objective is to find a safe linear description of the normal moments at a point S in the plate in relation to the yield criterion. Referring to Fig. 1 it can be shown that for any arbitrarily chosen ψ, where $\kappa = \tan\psi$ and is non-negative, at least one of M_A and M_B, which are linearised estimates of M_M, is greater than or equal to M_M.

3.1 Algebraic Expressions for Yield Criterion

3.1.1 Incremental Plastic Rotation (IC) Criterion

Adopting $\kappa^+ = \kappa^- = 1$, a simple but conventional criterion is obtained:

$$\begin{aligned}
R_{11} + R_{12} + \lambda F_1^{11} &\leqslant Q_{11}^+ + Q_{12}^+ && \text{(a)}\\
R_{22} + R_{12} + \lambda F_2^{11} &\leqslant Q_{22}^+ + Q_{12}^+ && \text{(b)}\\
R_{11} - R_{12} + \lambda F_1^{12} &\leqslant Q_{11}^+ - Q_{12}^+ && \text{(c)}\\
R_{22} - R_{12} + \lambda F_2^{12} &\leqslant Q_{22}^+ - Q_{12}^+ && \text{(d)}\\
- R_{11} + R_{12} + \lambda F_1^{21} &\leqslant - Q_{11}^- + Q_{12}^- && \text{(e)}\\
- R_{22} + R_{12} + \lambda F_2^{21} &\leqslant - Q_{22}^- + Q_{12}^- && \text{(f)}\\
- R_{11} - R_{12} + \lambda F_1^{22} &\leqslant - Q_{11}^- - Q_{12}^- && \text{(g)}\\
- R_{22} - R_{12} + \lambda F_2^{22} &\leqslant - Q_{22}^- - Q_{12}^- && \text{(h)}
\end{aligned} \qquad [4]$$

where $F_i^{k\ell} = \underset{t_\gamma,\ 0 \leqslant t_\gamma < \tau}{\text{Maximum}} [\delta_k E_{ii}^\gamma + \delta_\ell E_{ij}^\gamma]$; $i,k,\ell = 1,2$ and $j = 3-i$; $\delta_1 = 1$, $\delta_2 = -1$, and E_{ij}^γ are the components of state of moments $\{E^\gamma\}$ obtained from an initially stress free plate of unlimited elastic response under the loading $\frac{1}{\lambda}\mathbf{p}(t_\gamma)$ which gives

$$M_{ij}^\gamma = \lambda E_{ij}^\gamma + R_{ij}; \quad i,j = 1,2 \ . \qquad [5]$$

3.1.2 Alternating Plastic Rotation (AP) Criterion

By a similar argument,[10] defining T_{ij} as a component of moment variability capacity, we obtain the AP criterion in the form

$$\lambda \leqslant \frac{T_{ii} + \delta_k \, T_{ij}}{F_i^{1k} + F_i^{2\ell}} \qquad [6]$$

with the terms defined after Eqn. [4].

4. NUMERICAL EVALUATION

In the numerical method, the shakedown limit is obtained as the maximum value of λ within the conditions set by [1], [4] and [6]. The discretization introduced is as follows:

(i) The shakedown criteria are imposed only at nodal points formed by a rectangular curvilinear net chosen on the plate.
(ii) The load program is discretised into a finite number of load cases $\mathbf{L}_w$; $w = 1, \ldots, m$ where $\mathbf{L}_w = \mathbf{p}(t_w)$; $t_w \in \Omega$, .
The maximum value of λ for [1] and [4] could be computed independently of that for [6], since λ is the only common unknown variable involved in both. The former gives the IC limit λ_{IC} and the latter gives the AP limit λ_{AP}.

4.1 Incremental Collapse Limit

Applying the equilibrium condition for residual stresses [1] in finite difference form at all nodal points with modifications to account for the boundary conditions, we get

$$\mathbf{A} \cdot \mathbf{r} = 0 \qquad [7]$$

where $\mathbf{A}$ is the coefficient matrix, $\mathbf{r}$ is the vector of residual moments. Each nodal point yields one equilibrium equation.

The yield criterion for IC [4], applies eight constraints per nodal point, yielding, for the whole structure

$$\mathbf{B} \cdot \lfloor \frac{\mathbf{r}}{\lambda} \rfloor \leq \mathbf{a} \qquad [8]$$

where **B** is the matrix of coefficients and **a** is a vector of moment capacities. The values of $F_i^{k\ell}$ needed for **B** are to be evaluated from an elastic analysis.

The MP has been reduced to a Linear Progam (LP) to maximize $\lambda = \lambda_{IC}$ subject to [7] and [8]. From studying the active constraints in the optimal solution it is possible to estimate the incremental collapse mechanism.

4.2 Alternating Plasticity Limit

By computing the value of λ_o for each point, where $\lambda \leq \lambda_o$ is the equivalent form of [6], the AP limit λ_{AP} can be obtained as the minimum of those λ_o values of all nodal points.

5. EXAMPLE: RECTANGULAR PLATE UNDER A MOVING PATCH LOAD

A computer program was written using the numerical method above. SAP IV was used for the elastic analysis. As an example, the results for a simply supported isotropic rectangular plate (2ℓ x 3ℓ) under a rectangular (0.1ℓ x 0.2ℓ) patch load of unit intensity moving along the longer centre line is presented. The finite element mesh used is as shown in Figure 2. Only half of the plate was modelled for analysis, using symmetry. The plate was assumed to be isotropic with the material and sectional properties at all nodal points as follows:
Poisson ratio = 0.3, $Q_{11}^{+} = Q_{22}^{+} = -Q_{11}^{-} = -Q_{22}^{-} = m_p$, $T_{11} = T_{22} = 2m_a$. and $Q_{12}^{+} = Q_{12}^{-} = T_{12} = 0$. (In computing $m_p = m_a = 1$ unit was used.)

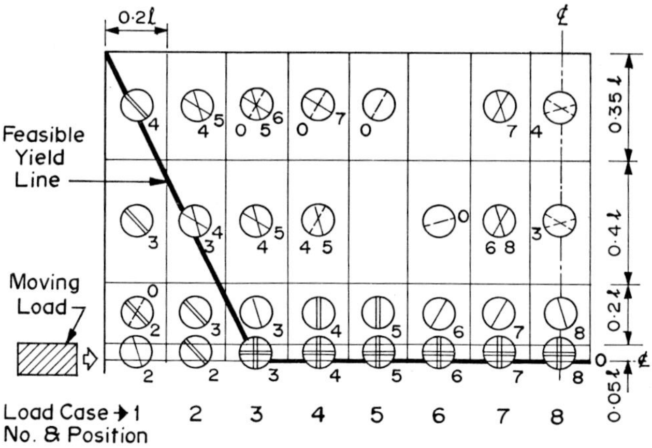

For explanation of plastic rotation symbols see References 10 & 13. Numbers shown with symbols correspond to load case numbers.

Fig. 2. Incremental Collapse of Simply Supported Plate, $2\ell \times 3\ell$, with Load moving along Centreline.

The resulting load factors for the case illustrated are as follows:

	Sandwich Plate	Solid Plate
	Form Factor = 1.0	Form Factor = 1.5
Elastic limit	$170.57\ m_a/\ell^2$	$170.57\ m_a/\ell^2$
	$= 170.57\ m_p/\ell^2$	$= 113.71\ m_p/\ell^2$
IC limit	$297.60\ m_p/\ell^2$	$297.60\ m_p/\ell^2$
AP limit	$341.15\ m_a/\ell^2$	$341.15\ m_a/\ell^2$
	$= 341.15\ m_p/\ell^2$	$= 227.43\ m_p/\ell^2$
Static collapse	$463.44\ m_p/\ell^2$	$463.44\ m_p/\ell^2$

It can be seen that IC controls the shakedown limit for the sandwich plate, and AP controls the shakedown limit for the solid plate.

6. EFFECT OF ANISOTROPIC MOMENT CAPACITY

The formulation for IC has been couched in terms which permit any relative values of the orthotropic capacities Q_x^+, Q_x^-, Q_y^+, and Q_y^-.

8.

The results of a study reducing the ratio of Q^+ to Q^- from 1.0 to zero for a simply supported sandwich plate 2ℓ x 3ℓ are shown in Fig. 3, using a slightly different mesh and patch dimensions from Fig. 2.

The results are unusual in that λ_{IC} approaches zero as the value of Q^- approaches zero when the dead load is zero. A study of the residual moment of equilibrium equation [1] reveals that, coupled with zero load as an active load case and the linear moment curvature relationship as described in the inset to Fig. 5, non-zero residual moments are impossible when $Q^- = 0$.

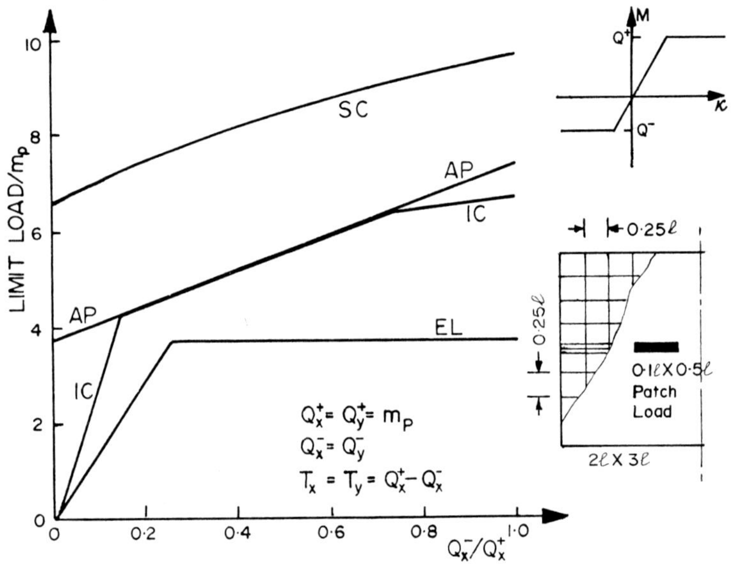

Fig. 3. Effect of Q^-/Q^+ Ratio on Capacity of a Sandwich Plate.

It is evident from this single instance that the linear model of elastic behaviour can be a severely restrictive one in practical terms.

9.

7. UPPER BOUND SOLUTION FOR IC

Knowledge of the likely IC mechanism can be used to circumvent the MP exercise entirely, although upper bound MP solutions for static collapse have been found.[11]

7.1 Upper Bound Formulation for a Plate

Expressing the general theorem in terms of stress resultants for a plate, incremental collapse is assured for any arbitrarily chosen yield line mechanism[3] when

$$\lambda\left[\int_{s^+} E_n^+\,\theta^+ ds + \int_{s^-} E_n^-\,\theta^- ds\right] \geqslant \int_{s^+} Q_n^+\,\theta^+ ds + \int_{s^-} Q_n^-\,\theta^- ds \qquad [9]$$

where s is the locus co-ordinate of the yield line, E_n is the maximum repeated moment normal to the yield line assuming unlimited elastic response in the total time Ω,
θ is the plastic rotation of the yield line, and + and - indicate the sign of plastic rotation. To avoid MP a less rigorous approach involves the use of a limited number of trial mechanisms, and selecting the lowest λ_{IC} from these trials. This leads to a double economy in computational effort. Firstly, there is no MP phase, and secondly, the elastic analysis requires results only along proposed yield lines. Although the approach is less rigorous, it is satisfactory when the incremental collapse mechanism is known within reasonable limits.

8. EXAMPLE: UPPER BOUND ANALYSIS OF RECTANGULAR PLATE.

The plate $2\ell \times 3\ell$ has been analysed by the upper bound method using simple trial collapse mechanisms. Stress resultants along the diagonal yield lines from the corners of the plate need to be properly evaluated. In using the finite element method for calculating these stress resultants, the mesh in the vicinity of these yield lines must not be too coarse if accuracy is to be achieved.

The results are shown in Fig. 4, for the isotropic slab analysed by the lower bound method in Fig. 2. The sensitivity of the result to the ratio a/ℓ, which defines the yield line geometry is evident. One advantage of the upper bound approach is that the exact moment capacity normal to the yield line can be used, rather than the linearised yield criteria used in the lower bound approach.

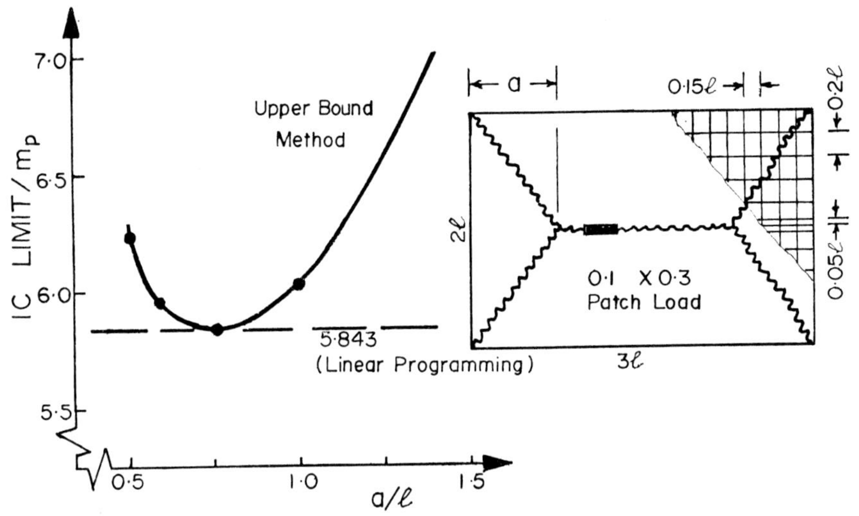

Fig. 4. Results of Upper Bound Analysis of Isotropic Plate 2ℓ x 3ℓ.

The coincidence of upper and lower bound solutions at a/ℓ = 0.75, when patch load is 5.843 m_p is not a guarantee of the exactness of the solution, since discretisation relaxes the strictness of the bounds on solution.

9. OBSERVATIONS ON THE METHODS OF ANALYSIS

The methods of analysis presented have proved practical for the estimation of the shakedown capacity of plates. The load cases used to demonstrate the techniques were somewhat specialised, but a more complex set of alternative load cases merely extends the elastic analysis by a small amount, and the MP phase is unaffected.

Many of the special and significant aspects of the behaviour of laterally loaded plates have been ignored in the development so far. For example, the arching action characteristic of reinforced concrete slabs, the membrane effects due to large deflections of thin slabs - in fact all membrane effects in both elastic behaviour and yield criterion - have been ignored. It is felt that the introduction of a yield criterion involving in-plane forces, which has been developed for reinforced concrete,[12] would need to be matched by an elastic analysis using appropriate nonlinear moment-curvature relationships as develop with cracked reinforced concrete.

10. REFERENCES

1. NATO Advanced Study Institute, University of Waterloo, 1977, Engineering Plasticity by Mathematical Programming. Cohn, M.Z. and Maier, G. (eds.), Pergamon, 1979.

2. MELAN, E. Der Spannungszustand eines Hencky-Mises'schen Kontinuums bei verandlicher Belastung. Sber. Akad. Wiss, Wien, 147, 73, 1938.

3. KOITER, W.T., A new general theorem on shakedown of elastic-plastic structures. Proc. K. ned. Akad. Wet., B59, 24, 1956.

4. KONIG, J.A. Theory of shakedown of elastic-plastic structures. Archwm. Mech. Stosow, 18, pp. 227-236, 1966.

5. GRUNDY, P. and SPENCER, W.J., Shakedown of Elastoplastic Grids. Proc. 5th Australasian Conf.Mechanics of Structures and Materials Melbourne, Aug. 1975, pp. 219-236.

6. KONIG, J.A. Shakedown of Plates. Archwm. Mech. Stosow, 21, pp. 623-637, 1969.

7. SAWCZUK, A. On incremental collapse of shells under cyclic loading. IUTAM Symposium, Copenhagen, 1967. Theory of Thin Shells, pp. 328-340, Springer-Verlag, 1969.

8. SAVE, M. A consistent limit analysis of reinforced concrete slabs. Mag. Conc. Res., V. 19, No. 58, pp. 3-12, 1967.

9. LENSCHOW, R. and SOZEN, M. A yield criterion for reinforced concrete slabs. J.A.C.I., Proc., V64, No. 5, pp. 266-273, 1967.

10. ALWIS, W.A.M. and GRUNDY, P., Shakedown of Plates under moving loads. Proc. 7th Austral. Conf. on Mechanics of Structures and Materials, Perth, May, 1980, pp. 191-196.

11. MUNRO, J. and Da FONSECA, A.M.A., Yield line methods by finite elements and linear programming. The Struct. Engineer, Vol. 56B, No. 2, pp. 37-44, June 1978.

12. COOKSON, P.J., Generalized Yield Lines in Reinforced Concrete Slabs. J. Struct. Mech., 7(1), pp 65-82, 1979.

13. GRUNDY, P. and ALWIS, W.A.M., Shakedown of Plates under Moving Patch Load. Monash Univ. Civ. Engg. Report No. 4/1980.

11. NOTATION

$\mathbf{a}$	vector of moment capacities
$\mathbf{A},\mathbf{B}$	coefficient matrices
E^{γ}	elastic moment due to load $\mathbf{p}(t_{\gamma})$
$F_i^{k\ell}$	envelope values of elastic moments; $i,k,\ell = 1,2$
$\mathbf{L}_w$	load vector of a load case
M^{γ}	moment due to load $\mathbf{p}(t_{\gamma})$
$\mathbf{p}(t)$	load vector
$\mathbf{q}(t)$	vector of time functions of load program
Q^+,Q^-	moment capacities
R	residual moment
t	time
T	moment variability capacity
$\mathbf{W}$	matrix of independent load vectors
x_i	Cartesian axies: $i = 1,2,3$
α,θ	angle of orientation measured anticlockwise from x_1 axis
ζ,η	principal orthotropic directions
λ	load multiplier
τ	duration of load program
Ω	set of time values in the domain $0 \leqslant t \leqslant \tau$

Abbreviations

MP	Mathematical Programming	LP	Linear Programming
EP	Engineering Plasticity	IC	Incremental Collapse
AP	Alternating Plasticity	SC	Static Collapse
EL	Elastic Limit.		

Comportement Asymptotique des Structures Elastoplastiques sous Chargement Periodique

B. HALPHEN
Laboratoire Central des Ponts et Chaussées,
Paris, France

NGUYEN Q.S.
Laboratoire de Mécanique des Solides,
Ecole Polytechnique, Palaiseau, France

Résumé : On rappelle les résultats relatifs à l'existence d'une solution périodique en contraintes ou en contraintes et déformations, et à la convergence de toute solution vers une solution périodique, dans les structures constituées de certains matériaux élastoplastiques soumises à un chargement périodique. On propose des méthodes de calcul permettant de construire rapidement une telle solution périodique.

1. Introduction

Certaines structures industrielles sont soumises à des charges et des distributions de température variant périodiquement dans le temps. Des raisons économiques et technologiques amènent parfois à accepter qu'elles subissent localement des déformations plastiques cycliques, lorsque le nombre de cycles de chargement est suffisamment faible.

L'objet de notre intervention est d'une part de rappeler les principaux résultats relatifs au comportement asymptotique des structures élastoplastiques soumises à un chargement thermomécanique périodique, et d'autre part de proposer des algorithmes de calcul numérique permettant d'obtenir la forme de la réponse asymptotique des systèmes étudiés plus rapidement qu'en déterminant toute leur évolution à partir de leur état initial.

2. Matériaux élastoplastiques standards généralisés de classe $\mathcal{C}$

Les structures étudiées ici sont constituées de matériaux élastoplastiques standards généralisés de classe $\mathcal{C}$ |1|,|2|. Leur vitesse de déformation totale $\mathbf{d}$, somme d'une vitesse de déformation thermoélastique $\mathbf{d}^e$ et d'une vitesse de déformation plastique $\mathbf{d}^p$, est donnée par :

$$\mathbf{d} = \mathbf{d}^e + \mathbf{d}^p = \Lambda\dot{\sigma} + B\dot{T} + \mathbf{d}^p \tag{1}$$

où σ, T désignent le tenseur de contraintes de Cauchy et la température absolue. La matrice Λ, supposée définie positive, et la matrice B sont constantes pour une particule matérielle donnée.

De plus, pour ces matériaux, il existe une famille $\alpha = \{\alpha_i\ ,\ i = 1,\dots,n\}$ de paramètres d'écrouissage et une matrice Z symétrique définie positive, constante pour une particule donnée, telles que si :

$$A = Z\alpha \tag{2}$$

la vitesse de déformation plastique et le taux des paramètres d'écrouissage vérifient :

$$(\mathbf{d}^p, -\dot{\alpha}) \in \partial\psi_{C(T)}(\sigma, A) \tag{3}$$

où $\partial\psi_{C(T)}$ désigne le sous-différentiel de la fonction indicatrice $\psi_{C(T)}$ d'un domaine convexe fermé $C(T)$ de l'espace $\sigma \times A$, appelé domaine d'élasticité du matériau dans cet espace. Cette relation (3) signifie que $(\mathbf{d}^p, -\dot{\alpha})$ représenté par un vecteur à 9 + n composantes, est une normale extérieure au domaine $C(T)$ au point (σ, A).

Les matériaux standards généralisés sont des cas particuliers de matériaux élastoplastiques standards écrouissables. En particulier on peut toujours trouver une extension tridimensionnelle standard généralisée des modèles analogiques constitués de ressorts et de patins à seuil constant |3|.

3. Formulation du problème d'évolution d'un volume de matériau standard généralisé de classe $\mathcal{C}$.

Considérons un volume fini $\mathcal{V}$, constitué d'un matériau élastoplastique standard généralisé de classe $\mathcal{C}$, soumis à un trajet de chargement thermomécanique de type classique ou dépendant d'un nombre fini de paramètres. Soit $K(t)$ l'ensemble convexe des champs de contraintes généralisées $\Sigma = (\sigma, A)$ plastiquement admissibles et statiquement admissibles à l'instant t , c'est-à-dire tels que :

$$\forall x \in \mathcal{V}\ ,\quad (\sigma(x), A(x)) \in C(x, T(x,t)) \tag{4}$$

et que le champ de contraintes soit statiquement admissible avec les données en efforts à l'instant t .

Désignons par $\sigma^E(t)$ la réponse purement élastique en contraintes du volume $\mathcal{V}$ et posons :

$$\Sigma^E(t) = (\sigma^E(t), 0) \tag{5}$$

et définissons le produit scalaire de deux champs de contraintes généralisées $\Sigma = (\sigma, A)$ et $\Sigma^* = (\sigma^*, A^*)$ par :

$$\langle \Sigma, \Sigma^* \rangle = \frac{1}{2}\int_{\mathcal{V}} \sigma \Lambda \sigma^* \, dv + \frac{1}{2}\int_{\mathcal{V}} A Z^{-1} A^* \, dv \tag{6}$$

Moreau |4|, pour le matériau parfaitement plastique, et Nguyen Q.S. |1|, |5|, pour le matériau écrouissable, ont établi que la loi d'écoulement et d'écrouissage (3) et le principe des puissances virtuelles permettent de mettre le problème d'évolution du champ de contraintes généralisées dans les structures considérées ici sous la forme d'une équation différentielle multivoque :

$$\dot{\Sigma}(t) + \partial\psi_{K(t)}(\Sigma(t)) \ni \dot{\Sigma}^E(t) \tag{7}$$

où $\partial\psi_{K(t)}$ désigne le sous-différentiel, au sens du produit scalaire (6), de la fonction indicatrice $\psi_{K(t)}$ du convexe $K(t)$.

4. Cas d'un chargement périodique

4.1. Généralités

Lorsque, pour la structure étudiée, toutes les données en efforts, en déplacements et en températures sont périodiques de période P, on a :

$$\Sigma^E(t+P) = \Sigma^E(t) \quad \text{et} \quad K(t+P) = K(t) \tag{8}$$

On peut alors se demander si, d'une part, il existe une solution périodique de l'équation d'évolution (7) et si, d'autre part, une solution de cette équation correspondant à des conditions initiales quelconques tend vers une solution périodique lorsque le temps tend vers l'infini.

La réponse positive à ces questions, qui ne peut être donnée que pour certaines formes particulières du convexe $K(t)$, c'est-à-dire pour certaines formes du domaine d'élasticité du matériau constituant la structure, repose toujours sur la propriété fondamentale suivante : l'application f_t définie par :

$$\Sigma(t) \xrightarrow[\theta]{f_t} \Sigma(t+P) \tag{9}$$

est une contraction du convexe $K(t)$ dans lui-même.

4.2. Matériau parfaitement plastique standard

Considérons un volume $\mathcal{V}$ de matériau élastique parfaitement plastique standard soumis à un chargement thermomécanique périodique. On démontre |6|, |7| :

Théorème : Si le domaine d'élasticité du matériau est borné, ou borné dans l'espace des déviateurs de contraintes, il existe une solution périodique du problème d'évolution des contraintes. Si de plus ce domaine est strictement convexe, ou indépendant de la température, et si l'espace des champs de contraintes est de dimension finie [(1)], toute solution du problème d'évolution des con-

traintes tend vers une solution périodique.
A une solution périodique en contraintes, on peut associer une évolution périodique des vitesses de déformation, mais le processus de rochet reste possible.

4.3. Matériau élastoplastique standard à écrouissage cinématique

Le domaine d'élasticité d'un matériau élastoplastique standard à écrouissage cinématique est défini par :

$$\frac{1}{2}\,(s_{ij} - A_{ij})(s_{ij} - A_{ij}) - k^2(T) \leqslant 0 \tag{10}$$

où s désigne le déviateur des contraintes, la force généralisée A évoluant suivant la loi :

$$\dot{A} = Z\, d^p \tag{11}$$

où Z est une matrice constante définie positive.
On démontre |7|, |9| :

Théorème : Dans un volume $\mathcal{V}$ de matériau élastoplastique standard à écrouissage cinématique, soumis à un trajet de chargement thermomécanique périodique, il existe une solution périodique du problème d'évolution des contraintes et des déformations ; si la dimension de l'espace des contraintes est finie, toute solution du problème d'évolution de la structure tend vers une solution périodique en contraintes et déformations.

5. Cas d'un chargement alterné

Considérons une structure constituée d'un matériau élastoplastique standard généralisé dont le domaine d'élasticité ne dépend pas de la température et est symétrique

---

(1) Johnson, Mercier et Nédélec |8| ont établi un résultat de convergence faible en dimension infinie.

par rapport à l'origine dans l'espace $\sigma \times A$, soumise à un chargement alterné, c'est-à-dire tel que :

$$\Sigma^E(t+\frac{P}{2}) = -\Sigma^E(t) \qquad (12)$$

On montre que, pour une solution du problème d'évolution de cette structure, l'application g_t définie par :

$$\Sigma(t) \xrightarrow[\theta]{g_t} -\Sigma(t+\frac{P}{2}) \qquad (13)$$

est une contraction du convexe $K(t)$ dans lui-même.
Pour les matériaux envisagés au paragraphe précédent, on démontre alors qu'il existe une unique solution alternée en contraintes généralisées de ce problème, et que toute solution tend vers une solution alternée autour d'une valeur moyenne éventuellement non nulle.
Si le matériau est parfaitement plastique, on peut associer à une solution en contraintes alternée autour d'une valeur moyenne non nulle une solution alternée en vitesses de déformation plastique, donc périodique en déformations.
Si le matériau est à écrouissage cinématique, on montre que les mêmes résultats subsistent si le chargement est lui-même alterné autour d'une valeur moyenne non nulle.

6. Calcul d'une solution périodique

6.1. Généralités

Dans les structures élastoplastiques soumises à un chargement périodique étudiées aux paragraphes 4 et 5, il existe généralement plusieurs solutions périodiques en contraintes généralisées, mais on montre que la différence entre deux solutions périodiques est constante dans le temps.
Le comportement asymptotique de ces structures dépend des conditions initiales, généralement mal connues, qui y règnent ; cependant la connaissance d'une solution périodique particulière fournit la forme de leur réponse asymptotique, et en particulier l'amplitude de variation des contraintes, ou des contraintes et des déformations.

6.2. Calcul d'une solution alternée

Lorsque le chargement du système étudié est alterné, et que les conditions du paragraphe 5 sont remplies, la détermination de la solution alternée du problème d'évolution des contraintes généralisées est équivalente à celle du point fixe de l'application g_0 qui associe $-\Sigma(\frac{P}{2})$ à $\Sigma(0)$. Pour obtenir celui-ci, on propose d'utiliser la suite de Picard définie par :

$$\Sigma_{n+1} = \omega \Sigma_n + (1-\omega)\, g_0(\Sigma_n) \quad , \quad 0 < \omega < 1 \tag{14}$$

qui converge vers ce point fixe.

On a traité de cette manière |10| l'exemple d'un tube encastré à ses deux extrémités, soumis à une répartition transversale de température alternée. Les résultats du calcul, avec $\omega = \frac{1}{2}$, qui utilise l'algorithme mis au point par Nguyen Q.S |5| pour passer de Σ_n à $g_0(\Sigma_n)$, montrent que la convergence de la suite (14) est deux fois plus rapide que celle d'un calcul incrémental complet.

6.3. Calcul d'une solution périodique

Lorsque le chargement de la structure étudiée est simplement périodique, le calcul d'une solution périodique en contraintes généralisées revient à celui d'un point fixe de l'application f_0 définie par (9). Comme la convergence vers une solution périodique de toute solution du problème d'évolution est généralement monotone, un processus d'interpolation tel que (14) ne ferait que retarder la convergence. On propose donc, pour déterminer un point fixe de f_0 d'utiliser la suite définie par :

$$\Sigma_{n+1} = f_0\, \mathcal{P}\left[\omega f_0(\Sigma_n) + (1-\omega)\Sigma_n\right] \quad , \quad \omega > 1 \tag{15}$$

où $\mathcal{P}$ désigne la projection sur $K(0)$. La convergence de cette suite n'a pas été démontrée ; mais dans le calcul d'un tube soumis à une pression interne constante et à une répartition transversale de température périodique |10|, on a constaté que la suite (15) converge vers une solution périodique plus rapidement qu'un calcul incrémental, même pour des valeurs grandes du paramètre ω (jusqu'à $\omega = 4$). La solution périodique trouvée dépend de Σ_0 et de ω .

Références

|1| Nguyen Q.S : Contribution à la théorie macroscopique de l'élastoplasticité avec écrouissage, Thèse, Paris 1973.

|2| B. Halphen et Nguyen Q.S : Sur les matériaux standards généralisés, J. de Mécanique, 14, n° 1, pp 39-63, 1975.

|3| B. Halphen et Nguyen Q.S : Sur les lois d'évolution des paramètres d'écrouissage, Cahiers du Gr. Fr. de Rhéologie, 3, n° 6, pp 247-250, 1975.

|4| J.J. Moreau : Systèmes élastoplastiques de liberté finie, Séminaire d'Analyse Convexe, Montpellier, 1973.

|5| Nguyen Q.S : On the elastic-plastic initial boundary value problem and its numerical integration, Int. Journal for Numerical Methods in Engineering, 11, pp 817-832, 1977.

|6| B. Halphen : Stress accommodation in elastic perfectly plastic and viscoplastic structures, Mec. Res. Comm, 2, N° 5 - 6, pp 273-278, 1976.

|7| B. Halphen : Accommodation et Adaptation des structures élastoviscoplastiques et plastiques, Matériaux et Structures sous chargement cyclique, Palaiseau 1978.

|8| C. Johnson, B. Mercier, J.C. Nédélec : Convergence vers une solution périodique en élastoplasticité parfaite, Matériaux et Structures sous chargement cyclique, Palaiseau 1978.

|9| B. Halphen : L'accommodation des structures élastoplastiques à écrouissage cinématique, C.R.Ac. Sc. Paris, t. 283, A, pp 799-802, 1976.

|10| M. Khouadja : Détermination de la réponse asymptotique des structures élastoplastiques et viscoplastiques sous chargements cycliques. Microthèse Ecole Polytechnique sous la direction de Nguyen Q.S, 1979 (non publiée).

A Constitutive Modelling of Time-Dependent Plastic Material with Work-Hardening and Recovery Effects

T. INOUE, S. NAGAKI, and S. TABATA

Department of Mechanical Engineering
Kyoto University
Yoshida-honmachi, Sakyo-ku 606, Kyoto, Japan

Summary

Kinematic hardening model is employed to describe the time-dependent constitutive relation by introducing the effects of recovery as well as hardening. Material parameters appearing in the evolutional equation for the center of yield surface are experimentally determined. Effect of deformation rate on tensile stress-strain diagram and creep curves are numerically calculated, and the results were compared with experimental data.

1. Introduction

To account for the time-dependent plastic behavior, which is observed for most engineering metals at elevated temperature, simple models of superposition of rate-independent plasticity and creep theory have been used. The authors [1] pointed out that such treaties neglecting the interaction effect between plastic and creep deformation are not available from the physical view-point as well as engineering sense. The purpose of the paper is to propose a unified constitutive model of inelastic behavior of metals at elevated temperature when both plastic hardening and the recovery are predominant.

2. Generalized kinematic hardening model

Recovery effects of metals at elevated temperature may be considered phenomenologically to originate from the disappearance of Bauschinger effect. When the pre-strained material is kept in stress free state, Bauschinger effect will tend to vanish as time goes on. In rate independent theory of plasticity, the effect is taken into account, in the first approximation, by use of kinematic hardening model of a yield surface. Prager's model assumes the form of the evolutional equation for the center of the yield surface $\underset{\sim}{\alpha}$,

$$\dot{\underset{\sim}{\alpha}} = C\,\dot{\underset{\sim}{\varepsilon}}^{p} \tag{1}$$

Here $\dot{\underset{\sim}{\varepsilon}}^{p}$ is inelastic strain rate and C is a function of equivalent plastic strain $\bar{\varepsilon}^{p}$ and temperature T.

Recovery effect occuring in once hardened material may be considerable if we introduce the modified relation

$$\dot{\underset{\sim}{\alpha}} = C_1(\bar{\varepsilon}^{p},\ T)\dot{\underset{\sim}{\varepsilon}}^{p} - C_2(\bar{\varepsilon}^{p},\ T)\underset{\sim}{\alpha} \tag{2}$$

Consider the unloaded state for the pre-strained material when stress is kept to be zero, the rate of change of the center is

$$\dot{\underset{\sim}{\alpha}} = -C_2(\bar{\varepsilon}^{p},\ T)\underset{\sim}{\alpha} \tag{3}$$

and integrating the equation leads to

$$\underset{\sim}{\alpha}(t) = \underset{\sim}{\alpha}_o\ exp\ \{-C_2(t - t_o)\} \tag{4}$$

where $\underset{\sim}{\alpha}_o$ is unloaded position of the center at time t_o. Equation (4) shows that the yield surface moves back to the initial position since $\underset{\sim}{\alpha}$ tends to zero when time elapses so far as C_2 is positive. It means that the recovery effect is adequately considerable due to the disappearance of Bauschinger effect. The first term of the right hand side of Eq. (2) is understood to represent strain hardening and the second term recovery effects. Similar kinds of idealization were recently made by Chang [2], Kujawski and Mroz [3]. The present consideration is, however, directly related to experimental observations.

To go further, we assume the flow rule as the form

$$\dot{\underset{\sim}{\varepsilon}}^{p} = \Lambda \frac{\partial F}{\partial \underset{\sim}{\sigma}} \tag{5}$$

where F is the yield function. Utilizing the consistency relation $\dot{F}$=0, parameter Λ is determined and Eq. (5) takes the form

$$\dot{\underset{\sim}{\varepsilon}}^{p} = \frac{\dfrac{\partial F}{\partial \underset{\sim}{\sigma}} : (\dot{\underset{\sim}{\sigma}} + C_2\underset{\sim}{\alpha})}{C_1 \dfrac{\partial F}{\partial \underset{\sim}{\sigma}} : \dfrac{\partial F}{\partial \underset{\sim}{\sigma}}} \frac{\partial F}{\partial \underset{\sim}{\sigma}} \tag{6}$$

In view of Eq. (6), the inelastic strain is not a homogeneous function of order one with respect to time rates of stress $\dot{\underset{\sim}{\sigma}}$ and temperature $\dot{T}$, which means that this is available to describe time dependence of plastic deformation.

3. Experimental and calculated results

The material parameters C_1 and C_2 appearing in Eq. (2) were determined for the low-carbon steel commercially called STB-35 from two kinds of experiments using thin-walled cylindrical specimen.

To know the effects of recovery, we measured the Bauschinger stress change in stress free state after pre-strained in torsion. Figure 1 shows variation of center of yield surface with time elapsed and function C_2 can be determined from the slopes of the lines in the figure as

$$C_2(\bar{\varepsilon}^p) = 6.52 \times 10^{-3}/\bar{\varepsilon}^p \tag{7}$$

When we use Mises type yield function F

$$F = \frac{1}{2}(\underset{\sim}{s} - \underset{\sim}{\alpha}) : (\underset{\sim}{s} - \underset{\sim}{\alpha}) - \frac{1}{3}\kappa^2 \tag{8}$$

the material parameter C_1 may be approximated as the slope of the stress-plastic strain curve at high stress rate, where $\underset{\sim}{s}$ is stress deviator and κ is constant, and we have

$$C_1(\bar{\varepsilon}^p) = 3.58\ (\bar{\varepsilon}^p + 0.15)^{-0.695} \tag{9}$$

When applying this model, material response for any loading programs may be calculated. Rate effects on stress-strain diagram are illustrated in Fig. 2, where solid lines are experimental results and broken lines are calculated ones through Eqs. (2)-(9). Figure 3

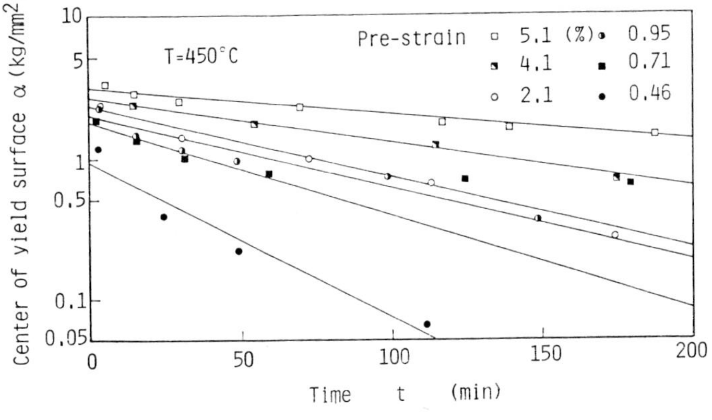

Fig. 1 Variation of center of yield surface.

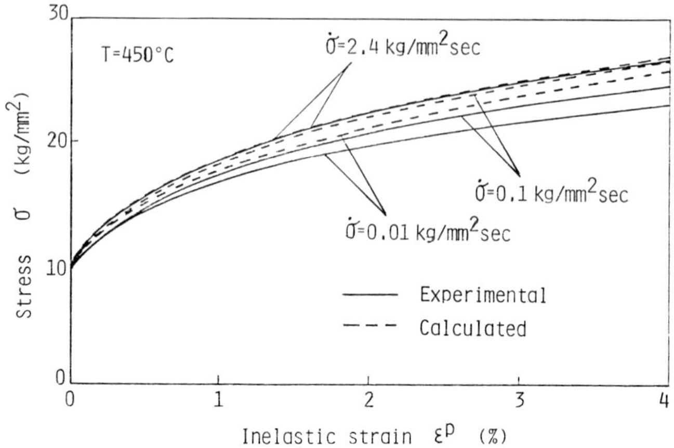

Fig. 2 Stress-strain curves at various stress rates.

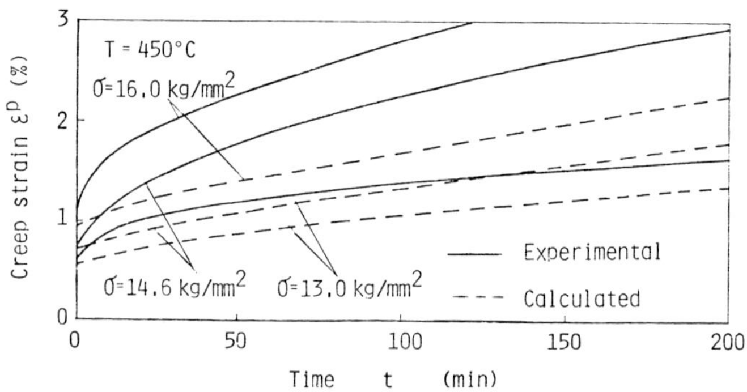

Fig. 3 Creep curves

shows results of static creep strain versus time. Calculated results (solid lines) agree fairly with experimental ones, especially in the region of stational creep.

References

[1] Inoue, T.; Nagaki, S.: A constitutive modeling of thermovisco-elastic-plastic materials. J. Thermal stresses, 1 (1978) 53-61.

[2] Chang, S.-J.: A creep-recovery constitutive equation and its time-independent limit. Proc. 9th SECTAM, Developments in Thoretical and Appl. Mech., 9 (1978) 497-508.

[3] Kujawski, D.; Mroz, Z.: A viscoplastic material model and its application to cyclic loading. Acta Mech., (1980) 243-252.

Limit Analysis of Prefabricated Large Panel Buildings by the Use of Discrete Models

S. KALISZKY

Department of Civil Engineering
Technical University, Budapest
Budapest 1521 Müegyetem rkp.3. Hungary

Summary

Paper presents simple discrete models which can be used for the limit analysis of prefabricated large panel buildings resting on a plastic subgrade. The model of the structure consists of rigid panels interconnected along their edges by plastic springs, while the subgrade is subdivided into prismatical elements which transmit normal, shear and couple forces to each other. Using these discrete models and linear/ized/ yield conditions statically admissible stress fields can be constructed for the structure and the subgrade. This makes possible the unified limit analysis of the whole system which leads to linear programming.

1. The model of the structure

Since the joints made in situ are generally the weakest parts of a prefabricated structure, in the present analysis of large panel /LP/ buildings we assume, that in plastic limit state merely the joints of the panels undergo plastic deformations. Thus, in the model of the LP structure the panels can be considered rigid elements, while the joints are replaced by plastic springs acting in tension, compression and in shear /Fig.1./ [1]. Considering a wall of a LP building the equilibrium of the i th rigid element /Fig.2.a/ subjected to the external loads and spring forces

$$\underline{\underline{q}}_i^* = [P_{ix}\ P_{iy}\ M_i] , \quad \underline{\underline{s}}_{ik}^* = [N'_{ik}\ N''_{ik}\ T_{ik}]$$

can be expressed as

$$\underline{\underline{G}}_i^*\ \underline{\underline{s}}_i + \underline{\underline{q}}_i = \underline{\underline{0}} \ . \qquad /1/$$

Here k = 1,2,3,4 refers to the edges of the element and

$$\underline{\underline{G}}_i^* = \left[\begin{array}{ccc|ccc|ccc|ccc} 0 & 0 & -1 & 1 & 1 & 0 & 0 & 0 & 1 & -1 & -1 & 0 \\ 1 & 1 & 0 & 0 & 0 & 1 & -1 & -1 & 0 & 0 & 0 & -1 \\ -b_i & b_i & a_i & a_i & -a_i & b_i & b_i & -b_i & a_i & -a_i & a_i & b_i \end{array}\right], \quad \underline{\underline{s}}_i = \begin{bmatrix} \underline{\underline{s}}_{i1} \\ \underline{\underline{s}}_{i2} \\ \underline{\underline{s}}_{i3} \\ \underline{\underline{s}}_{i4} \end{bmatrix}.$$

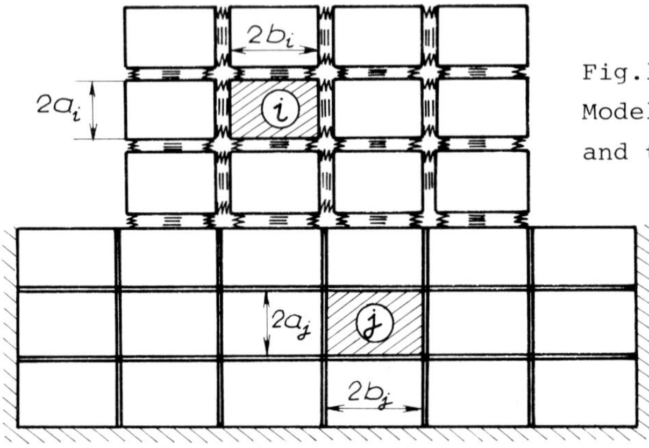

Fig.1.
Model of the structure and the subgrade

Denoting the yield forces of the springs in tension, compression and in shear by $\mathcal{N}_{ik}^+, \mathcal{N}_{ik}^-$ and $\mathcal{T}_{ik}$ and neglecting the interaction among the springs the yield conditions of the edges have the form

$$\underline{\underline{H}}_{ik}\underline{\underline{s}}_{ik} - \underline{\underline{r}}_{ik} \leq \underline{\underline{0}}; \qquad k = 1,2,3,4 \qquad /2/$$

Here

$$\underline{\underline{H}}_{ik}^* = \begin{bmatrix} 1 & -1 & 0 & 0 & 0 & 0 \\ 0 & 0 & 1 & -1 & 0 & 0 \\ 0 & 0 & 0 & 0 & 1 & -1 \end{bmatrix}, \quad \underline{\underline{r}}_{ik}^* = [\mathcal{N}_{ik}^+ \, \mathcal{N}_{ik}^- \, \mathcal{N}_{ik}^+ \, \mathcal{N}_{ik}^- \, \mathcal{T}_{ik} \, \mathcal{T}_{ik}].$$

2. The model of the subgrade

The ideal plastic subgrade /soil/ is subdivided into discrete prismatical elements which transmit normal, shear and couple forces to each other along their contact surfaces /Fig.1./[2]. Considering the j th element /Fig.2.b/ subjected to its own weight and to the contact forces

$$\underline{\underline{g}}_j^* = 4a_j b_j c\gamma_j [0 \ 1 \ 0], \qquad \underline{\underline{p}}_{jk}^* = [n_{jk} t_{jk} m_{jk}]$$

the equilibrium equations can be written as

$$\underline{\underline{B}}_j^{*}\ \underline{\underline{p}}_j + \underline{\underline{g}}_j = \underline{\underline{O}}. \qquad /3/$$

Here k = 1,2,3,4 refers to the boundary surfaces, c and γ_j denote the thickness and the specific weight of the element and

$$\underline{\underline{B}}_j^{*} = \left[\begin{array}{ccc|ccc|ccc|ccc} 0 & 1 & 0 & 1 & 0 & 0 & 0 & -1 & 0 & -1 & 0 & 0 \\ 1 & 0 & 1 & 0 & 1 & 0 & -1 & 0 & 0 & 0 & -1 & 0 \\ 0 & -a_j & 1 & 0 & b_j & -1 & 0 & -a_j & -1 & 0 & b_j & 1 \end{array}\right], \quad \underline{\underline{p}}_j = \begin{bmatrix} \underline{\underline{p}}_{j1} \\ \underline{\underline{p}}_{j2} \\ \underline{\underline{p}}_{j3} \\ \underline{\underline{p}}_{j4} \end{bmatrix}.$$

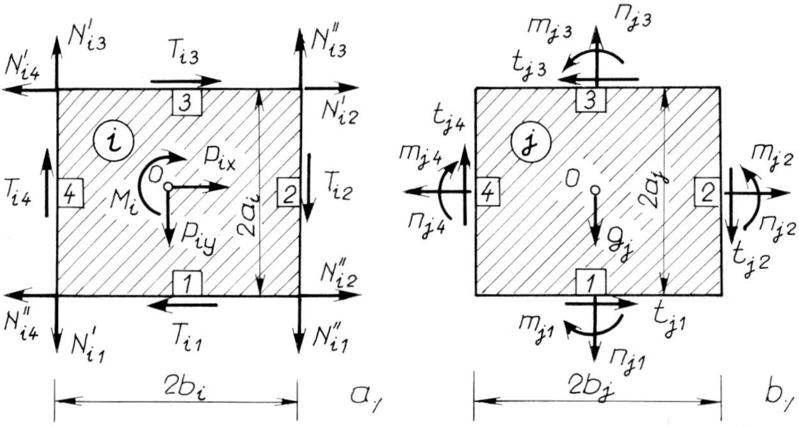

Fig.2. Discrete elements a/ structure, b/ subgrade

Applying the linearized Tresca-yield condition to the stress state of the middle point O of the element we obtain the inequalities

$$\underline{\underline{C}}_j\ \underline{\underline{p}}_j - \underline{\underline{\sigma}}_j \leq \underline{\underline{O}}. \qquad /4/$$

Here

$$\underline{\underline{C}}_j = \left[\begin{array}{ccc|ccc|ccc|ccc} -1 & \frac{3}{8} & \frac{9}{8a_j} & \frac{b_j}{a_j} & \frac{3b_j}{8a_j} & \frac{9}{8a_j} & -1 & \frac{3}{8} & -\frac{9}{8a_j} & \frac{b_j}{a_j} & \frac{3b_j}{8a_j} & -\frac{9}{8a_j} \\ -1 & -\frac{3}{8} & -\frac{9}{8a_j} & \frac{b_j}{a_j} & -\frac{3b_j}{8a_j} & -\frac{9}{8a_j} & -1 & -\frac{3}{8} & \frac{9}{8a_j} & \frac{b_j}{a_j} & -\frac{3b_j}{8a_j} & \frac{9}{8a_j} \\ 1 & -\frac{3}{8} & -\frac{9}{8a_j} & -\frac{b_j}{a_j} & -\frac{3b_j}{8a_j} & -\frac{9}{8a_j} & 1 & -\frac{3}{8} & \frac{9}{8a_j} & -\frac{b_j}{a_j} & -\frac{3b_j}{8a_j} & \frac{9}{8a_j} \\ 1 & \frac{3}{8} & \frac{9}{8a_j} & -\frac{b_j}{a_j} & \frac{3b_j}{8a_j} & \frac{9}{8a_j} & 1 & \frac{3}{8} & -\frac{9}{8a_j} & -\frac{b_j}{a_j} & \frac{3b_j}{8a_j} & -\frac{9}{8a_j} \end{array}\right]$$

$$\underline{\underline{\sigma}}_j^{*} = 4b_j c\sigma_j^{o}\ [1\ 1\ 1\ 1]$$

and σ_j^{o} denotes the yield stress of the j th element of the subgrade.

3. Limit analysis

Using the models described above statically admissible stress fields can be constructed for the LP structure and the subgrade which makes possible the unified limit analysis of the whole system. Considering a wall subjected to a one-parameter loading $\underline{\underline{q}} = m\underline{\underline{q}}_o$ the equilibrium and the yield conditions have the form

$$\underline{\underline{G}}^{*}\underline{\underline{s}} + m\underline{\underline{q}}_o = \underline{\underline{0}}, \qquad \underline{\underline{H}}\ \underline{\underline{s}} - \underline{\underline{r}} \leq \underline{\underline{0}}, \qquad /5/$$

while the same conditions for the subgrade are

$$\underline{\underline{B}}^{*}\underline{\underline{p}} + \underline{\underline{g}} = \underline{\underline{0}}, \qquad \underline{\underline{C}}\ \underline{\underline{p}} - \underline{\underline{\sigma}} \leq \underline{\underline{0}}. \qquad /6/$$

Since $\underline{\underline{s}}$ and $\underline{\underline{p}}$ are related by the statical boundary conditions of the contact line of the structure and the subgrade /5/ and /6/ construct a system of linear equations and inequalities the solution of which together with the extremal condition

$$m = \max! \qquad /7/$$

leads to linear programming.

4. Applications

The method described is suitable to the analysis of the structural response of LP buildings under abnormal loading conditions. Assuming ineffective /missing/ panels or subgrade elements the effect of gas explosion or of cavitation of the soil can be investigated. Besides, the discrete models can be applied to the approximate dynamic analysis of LP structures subjected to earthquake loading.

References

1 Kaliszky,S.; Wolf,K.: Analysis of panel buildings by the use of rigid panel model. Periodica Polytechnica Civil Eng. Vol.23. 3./1979/.

2 Kaliszky,S.: Statically admissible stress fields in plane plastic problems. Bul. Acad. Pol. Sci. Vol.XXVII. No.5/6. /1979/.

Inelastic Constitutive Equations and Phenomenological Laws of Damage Accumulation for Structural Metals

E. KREMPL

Department of Mechanical Engineering,
Aeronautical Engineering & Mechanics
Rensselaer Polytechnic Institute
Troy, New York 12181 U.S.A.

Summary

During inelastic deformation the mechanical properties of metals can change considerably. Examples of such changes are given. It is argued that these changes are not necessarily a reliable indicator of damage. Residual strength or residual lifetime are proposed as alternate relative measures. For the case of residual lifetime a method of assessing damage is proposed. This method yields the Palmgren-Miner rule as a special case.

1. Introduction

Modern engineering practice requires the prediction of the lifetime of severely stressed components in the early design stage. Traditionally lifetime calculations consist of a stress analysis part which furnishes the variation of the stress (strain) components as a function of time and location under a given operating history (boundary condition). The so-calculated stress and strains are then compared with allowable stresses (strains) and with creep (fatigue)-curves to assess the safety margin in monotonic loading and to obtain the lifetime of the component, respectively. This part is sometimes referred to as life-analysis and is, compared to stress analysis, an empirical procedure.

Failure of structural metal components is, with the exception of pure brittle fracture, almost always accompanied by inelastic* deformation which is highly nonlinear. With the advent of the computer nonlinear stress analyses can be and are being performed in the

---

*Inelasticity denotes any deviation from linear elasticity.

design office. The continuing development of modern constitutive equations, which are conceived to give a realistic representation of the rate(time)-dependent behavior of structural metals [1-10], and of the computational capabilities provide a means of accurately determining local stresses and strains as a function of time.

Presently most of the life analysis in engineering* uses "static concepts." A constant stress is compared with the creep rupture curve; at a given lifetime a stress(strain)-amplitude is shown to be below the permitted one, and a monotonically reached maximum stress falls below the yield limit or the allowable stress for buckling.

The use of finite element analysis with rate-dependent constitutive equations now yields time varying stresses (strains) at a given location. Under these circumstances the above described method of life-analysis is not commensurate with the accuracy of the computed stresses (strains). Essentially new methods for lifetime calculations must be developed and must be shown to be realistic and useful in design in the new design environment where modern computing power is available.

Until recently the subject of life-analysis was a stepchild of applied mechanics. This picture, however, is rapidly changing. "Continuous Damage Mechanics" [11] is being developed to meet the challenges of the future.

The purpose of this contribution is to comment on the interrelation between constitutive equations and damage accumulation laws and to propose residual lifetime as a measure of damage. Our attention will be focused on problems related to creep and fatigue (low-cycle and high-cycle) of structural metals in the context of a phenomenological, continuum mechanics approach.

---

*In some instances the cumulative effect of varying stresses (strains) is considered by using linear cycle (life, strain) fraction rules.

2. Mechanical property changes due to loading and environment.

The mechanical properties of structural metals can change as a consequence of exposure to environment in which case we speak of aging and due to prior mechanical loading defined as history dependence in the sense of plasticity (HISP)* [12].

The method of identifying these phenomena is to apply the same test condition (the input) to identically prepared samples which were exposed either to the environment alone (no mechanical loading) or to the environment and mechanical loading. By comparing the responses of the various samples aging and HISP can be definitely identified using this totally phenomenological method [12].

These phenomenologically identified changes are of course caused by changes in the microstructure of the material. In the case of aging the microstructural changes are caused by diffusion processes and chemical reactions. HISP can be caused by a change in dislocation density due to mechanical loading. Microstructural investigations using light and electron microscopy have documented these changes, see for example [13,14].

Figures 1 and 2 show examples of changes in the stress-strain behavior measured with AISI Type 304 Stainless Steel at room temperature. Figure 1 [15] demonstrates the well-known fact that the stress-strain curve of an annealed material can be substantially altered due to prior cold-working. In Figs. 2a and 2b the response of this material to step changes in strain rate is shown in the annealed condition and after cyclic loading. The increase in the stress level (flow stress) is evident. However, the difference between the flow stress at $10^{-3}\ s^{-1}$ and at $10^{-5}\ s^{-1}$ appears to be constant; see [16,17] for similar observations.

These room temperature examples may suffice to demonstrate that changes in mechanical properties due to prior mechanical loading

*Aging and HISP may also interact in the sense that mechanical loading initiates the aging process. Such interactions are present in strain aging and age hardening.

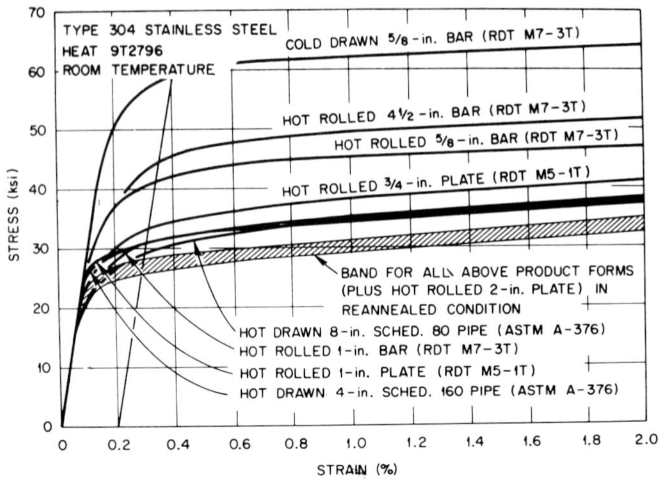

Fig.1. The influence of prior mechanical loading on the stress-strain diagram of AISI Type 304 Stainless Steel at room temperature [15]. It is seen that hot rolling and cold drawing (prior mechanical histories) increase the flow stress.

can be significant. (Further examples pertinent to creep and cyclic loading at elevated temperature are given in [18]).

The definitions of HISP and aging [12] can be applied to constitutive equations rather than real materials and it can be shown [12,18] that linear and nonlinear elasticity and linear and nonlinear viscoelasticity are incapable of reproducing HISP and aging*. Indeed, ideal materials which are incapable of a change in mechanical properties can be defined theoretically [19].

3. Identification of damage in real materials.

Damage denotes any injury or harm to one's person or estate and is synonymous with detriment, harm, injury or loss [20].

When we refer to damage in materials we want to maintain the same meaning of the word. The task is then to identify damage within the context of a phenomenological approach.

---

*If the kernels of the viscoelastic constitutive equations are not of the convolution form then aging can be reproduced.

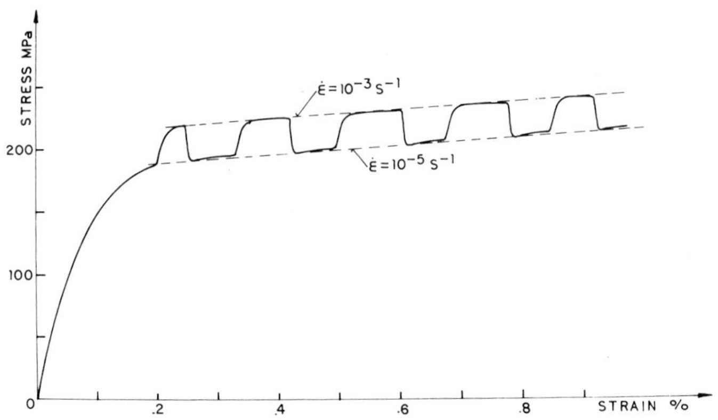

Fig.2a. The response of annealed AISI Type 304 SS to repeated instantaneous changes in strain rate at room temperature. It is seen that each strain rate has a characteristic stress level.

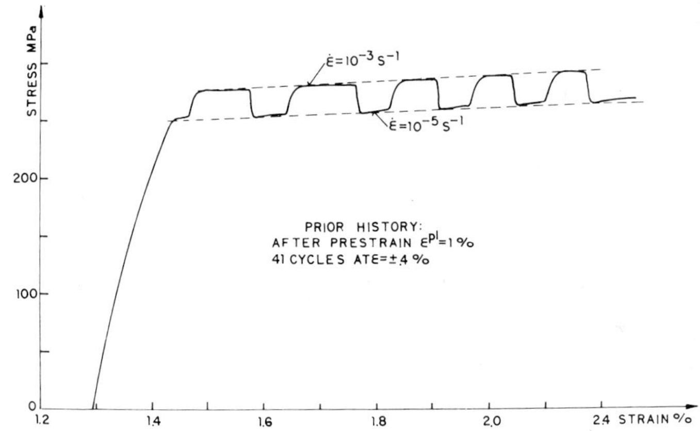

Fig.2b. Response of the same specimen to the same instantaneous changes in strain rate as in Fig. 2a but after monotonic straining and cycling. The stress level (flow stress) has increased considerably but the difference in the stress level at the two strain rates is the same as in Fig. 2a.

At the outset it can be stated that mechanical loading of sufficient severity and duration will ultimately cause failure. At failure damage is complete. To infer from this observation that every mechanical loading will bring the material closer to failure would be premature. Damage may not increase all the time. It is known that in some instances a certain threshold must be exceeded before damage can be done to a material. Constant amplitude stressing below the endurance limit (if it exists) will not cause failure. Further prior mechanical loading may indeed improve fatigue performance. Figure 6.15 of [21] shows that prior compression increases the endurance limit.*

It is therefore not possible to postulate that every mechanical loading will increase the "damage"; some mechanical loads may decrease "damage" and "healing" may occur.** "Damage" cannot always be monotonically increasing. Various means of identifying damage have been proposed which are based on the deformation behavior. They include (see [11] for additional references):

changes in elastic modulus [22]
changes in plastic strain range [23]
changes in stress range [23]
changes in the creep curve [23,24].

The examples given previously in this paper indicate that changes in deformation behavior are not necessarily a reliable indicator of damage. It is not known whether the fatigue or creep life of the annealed AISI Type 304 SS specimen (the mechanical response of which is given in Fig. 2a) is different from the preloaded AISI Type 304 SS specimen (mechanical response given in Fig. 2b). The increase of the endurance limit by prior compression cited previously [21, Fig. 6.15] is a fact which cannot easily be predicted from changes in deformation behavior (the stress-strain diagrams of the as received and of the precompressed specimens are certainly different; but how does one predict from this difference the increase in the endurance limit?).

It is known that changes in mechanical and physical properties occur during fatigue stressing (see [25, p.342]), but their exact relation to damage needs further clarification.

---

*The data refers to the endurance limit. It is possible that low-cycle fatigue life is decreased by cold working.

**See the example on the influence of prior compression on the fatigue limit [21], Fig. 6.15.

4. Alternate indirect damage measures.

An alternate indirect measure of damage is now proposed. It is felt that quantities associated with failure are more appropriate indicators of damage than quantities associated with deformation.

At failure damage is complete. It is therefore proposed to use phenomenological quantities measured at failure for the indirect determination of damage. They include residual lifetime and residual strength or residual deformation capability.

For a consistent evaluation failure must be defined. In accordance with generally accepted notions failure in this context is defined as crack initiation which has occurred when a crack is present whose length is comparable to the grain size of the metal (see [11]).

Residual lifetime and residual strength (deformation) are *relative* measures of damage. The lifetime or strength of a material is determined in the as received condition and after mechanical loading. Their comparison provides an unambiguous measure of the damage associated with prior mechanical loading.

5. Residual lifetime as a relative measure of damage.

Time is chosen rather than cycles to designate the event failure since creep and fatigue can easily be compared with this measure. In a periodic loading, lifetime can always be readily converted to cycles-to-failure by using the period of the cycle. Last and not least, the use of lifetime avoids the usual cycle counting difficulties in the case of random loading.

First we consider a set of *preloading histories* which may include creep, low-cycle fatigue, random loading, or any other loading history. We define the time ratio p_i by

$$p_i \equiv \frac{\text{Time spent in } i^{th} \text{ preloading history}}{\text{Time-to-failure for } i^{th} \text{ preloading history}}$$

or

$$p_i = \frac{t_i}{T_i} \; ; \quad 0 \leq p_i \leq 1 \qquad (1)$$

where the subscript i designates a given type of preload history.

Subsequently a test history α is applied. The time ratio $r_{i\alpha}$ is defined by

$$r_{i\alpha} \equiv \frac{\text{time-to-failure in } \alpha^{th} \text{ test history after } i^{th} \text{ preloading history}}{\text{time-to-failure in } \alpha^{th} \text{ test history without preloading history}} \qquad (2)$$

$$r_i = \frac{t_{i\alpha}}{T_\alpha} \geq 0$$

The time ratio $r_{i\alpha}$ is now used as a relative measure of damage.

For the zero prehistory (i=0) and the test history $\hat{\alpha}$, $r_{o\hat{\alpha}} = 1$. For any other prehistory say $\hat{i}$ and $p_{\hat{i}} > 0$, $r_{\hat{i}\hat{\alpha}}$ can deviate from one and this deviation is used as an indicator of damage. If $r_{\hat{i}\hat{\alpha}} < 1$ then the prehistory $\hat{i}$ applied during $p_{\hat{i}}$ was damaging for $\hat{\alpha}$; if $r_{\hat{i}\hat{\alpha}} > 1$ the opposite is true. It is essential to note that the statements are specific to a given combination of pre- and test-history. Even if $r_{\hat{i}\hat{\alpha}} < 1$ for the $\hat{\alpha}$-test history it is possible that $r_{\hat{i}\bar{\alpha}} > 1$ where $\bar{\alpha}$ denotes a test history different from $\hat{\alpha}$.

This possibility is indicated in Fig. 3 where p_i is plotted vs. $r_{i\alpha}$. Suppose that i designates a specific prehistory. All tests are interrupted at $p_i = \text{const}$. Subsequent tests with different test histories $\alpha = 1,2,\ldots 4$ yield for each α a specific $r_{i\alpha}$ as shown in Fig.3.

An extension of this thought experiment shows that for each combination of prehistory and test-history a separate curve on the p_i-$r_{i\alpha}$ plane is obtained. In general $f(p_i, r_{i\alpha}) = 0$ defines a failure curve for a given i-α pair with $f(1,0) = f(0,1) = 0$. One such hypothetical curve is completely drawn in Fig. 3, where $r_{i\alpha}$ values larger than one are shown to indicate that "healing" is possible.

For specific test conditions special forms of the failure loci $f(p_i, r_{i\alpha}) = 0$ can be deduced by theoretical arguments.

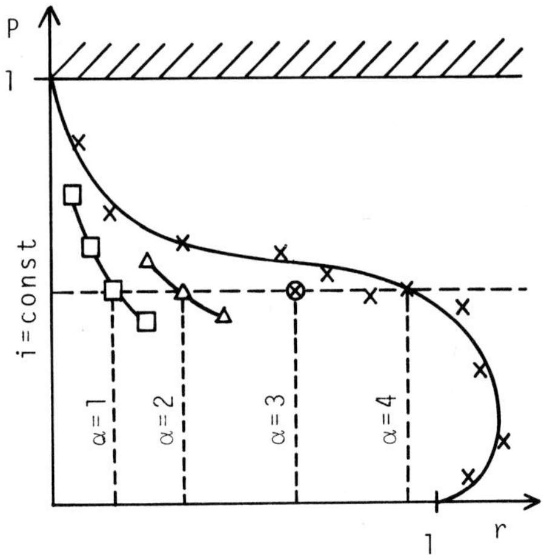

Fig.3. Schematic p-r diagram. A hypothetical prehistory is imposed and the test stopped at p = const. Subsequent tests with different test histories identified by different values of α, $\alpha = 1,2,3,4$ are performed. Each α-history will in general yield a different p-r curve as indicated.

The familiar Palmgren-Miner rule (which connects $p = 1$ with $r = 1$ by a straight line) is obtained if

- preload and test history are equal and aging* is absent
- the nature of preload and test-history has no influence on failure.

If no sequence effects** are observed then the failure loci are symmetric with respect to the bisector of the first quadrant of the p-r plane. This can be demonstrated as follows.

For a given i,α pair

$$p_i + r_{i\alpha} = a \tag{3}$$

where a is some number. Now we interchange the role of pre- and test-history to get when sequence effects are absent

$$P_\alpha + r_{\alpha i} = a \tag{4}$$

*If aging is important time spent between the end of the preloading history and the start of the test-history can have an influence.

**Sequence effects are absent if the interchange of pre- and test-history does not change the outcome of the experiment, see [26] for a description of this effect.

where the conditions of the second test are such that

$$p_\alpha = r_{i\alpha} \tag{5}$$

Comparing (3),(4) and (5) we must conclude that

$$p_i = r_{\alpha i} \tag{6}$$

Equations (3)-(6) imply that in the absence of sequence effects

$$f(p_i,\ r_{i\alpha}) = f(p_\alpha,\ r_{\alpha i}) = 0 \tag{7}$$

and the failure curves possess the symmetries shown in Fig. 4.

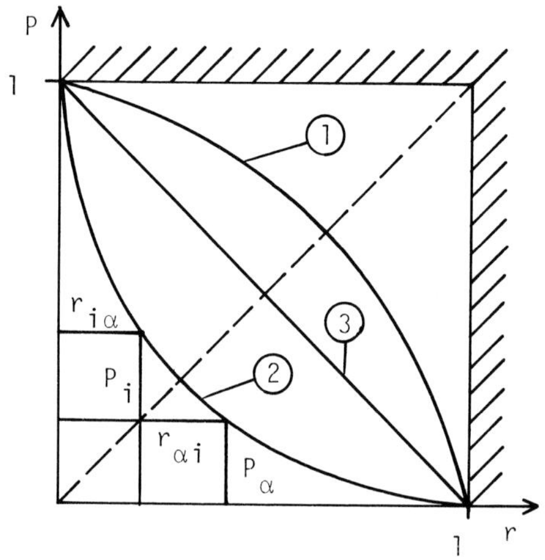

Fig.4. Schematic p-r diagrams in the absence of sequence effects. Note that the p-r curves are now bounded by the unit square. For 1 $p_i + r_{i\alpha} > 1$; for 2 $p_i + r_{i\alpha} < 1$. 3 represents the Palmgren-Miner rule for which $p_i + r_{i\alpha} = 1$.

The above arguments demonstrate that the Palmgren-Miner rule can only be expected to hold in special cases. In general a nonlinear p-r relation has to be expected. Indeed creep experiments reported in [27] show such a general dependence.

In [28] the existence of a unique state of damage is postulated by requiring that identical specimens subjected to different preloading histories will have the same residual lifetime for a given test-history α. This proposal does not require that the state of damage be characterized by a number. It is possible that different preloading histories which are terminated at the same p value can have different r values for the same α-history. In this proposal the relative measure of damage depends on the time ratios p and r and the nature of the pre- and test-history.

6. Damage Accumulation Laws.

The proposed relative measure of damage requires that the damage accumulation laws can represent the sensitivity of damage to the nature of pre- and test-history. The p-r diagrams can be used to ascertain qualitatively whether a given damage accumulation law can reproduce the different kinds of p-r curves. Such an attempt has been made in [26].

7. Acknowledgement.

This research was supported by the National Science Foundation and the Office of Naval Research.

8. Literature.

[1] Bodner, S.R.; Partom, Y.: Constitutive equations for elastic-viscoplastic strain hardening materials. J. Applied Mechanics, 42 (1975) 385.

[2] Eisenberg, M.A.; Lee, C.W.; Phillips, A.: Observations on the theoretical and experimental foundations of thermoplasticity. Int. J. Solids and Structures, 13 (1977) 1239-1255.

[3] Hart, E.W.: Constitutive equations for the non-elastic deformation of metals. Trans. ASME, J. Engineering Materials and Technology, 98H (1976) 193.

[4] Miller, A.K.: An inelastic constitutive model for monotonic, cyclic and creep deformation - Part I and II. Trans. ASME, J. Engineering Materials and Technology, 98H (1976) 97.

[5] Swearengen, J.C.; Rhode, R.W.: Application of mechanical state relations at low and high homologous temperatures. Met. Trans., 8A (1977) 577.

[6] Valanis, K.C.: On the foundations of the endochronic theory of viscoplasticity. Archives of Mechanics, 27 (1975) 857.

[7] Cernocky, E.P.; Krempl, E.: A theory of viscoplasticity based on infinitesimal total strain. Acta Mechanica, 36, No.3-4 (1980).

[8] Liu, M.C.M.; Krempl, E.: A uniaxial viscoplastic model based on total strain and overstress. J. Mechanics and Physics of Solids, 27 (1979) 377-391.

[9] Walker, K.P.; Krempl, E.: An implicit functional theory of viscoplasticity. Mechanics Research Communications, 4 (1978) 179-190.

[10] Krempl, E.: Viscoplasticity based on total strain. The modelling of creep with special considerations of initial strain and aging. Trans. ASME, J. Engineering Materials and Technology, 101 (1979) 380-386.

[11] Chaboche, J.L.: Continuous damage mechanics. A tool to describe phenomena before crack initiation. Paper presented at SMiRT-5 post-conference seminar on inelastic analysis and life-prediction in high temperature environment, Berlin (August 1979) to appear in Nuclear Eng. and Design.

[12] Krempl, E.: On the interaction of rate and history dependence in structural metals. Acta Mechanica 22 (1975) 53-90.

[13] Feltner, C.E.; Laird, C.: Cyclic stress-strain response of fcc metals and alloys. Parts I and II. Acta Metallurgica, 15 (1967) 1621-1653.

[14] Nahm, H.; Moteff, J.; Diercks, D.R.: Substructural development during low-cycle fatigue of AISI type 304 stainless steel at 649°C, Acta Metallurgica, 27 (1977) 107-116.

[15] RDT Standard RDT F 9-5T, Division of Reactor Research and Development, U.S. Atomic Energy Commission (Sept. 1974), Fig. A.35.

[16] Krempl, E.: An experimental study of room temperature rate-sensitivity, creep and relaxation of type 304 stainless steel. J. Mechanics and Physics of Solids, 27 (1979) 363-375.

[17] Kujawski, D.; Kallianpur, V.; Krempl, E.: Uniaxial creep, cyclic creep and relaxation of AISI type 304 stainless steel at room temperature. To appear J. Mechanics Physics of Solids.

[18] Krempl, E.: Cyclic creep. An interpretive literature survey. Welding Research Council Bulletin No. 195, Welding Research Council, New York, NY (1974).

[19] Stouffer, D.C.; Strauss, A.M.: A theory of material divagation. Int. J. of Eng. Science, 16 (1978) 1019-1028.

[20] Webster's New Twentieth Century Dictionary of the English Language, unabridged, 2nd edition, William Collins Publishers inc. (1979).

[21] Frost, N.E.; Marsh, K.J.; Pook, L.P.: Metal fatigue. Clarendon Press, Oxford (1974).

[22] Lemaitre, J.; Cordebois, J.-P.; Dufailly, J.: Sur le couplage endommagement-élasticité. C.R. Acad. Sc. Paris, t.288, Série B (25 Juin 1979) 391-394.

[23] Lemaitre, J.; Chaboche, J.-L.: Aspect phénoménologique de la rupture par endommagement. J. de Mecanique applicquée, 2 (1978) 317-365.

[24] Leckie, F.A.: The constitutive equations of continuum creep damage mechanics. Phil. Trans. R. Soc., London, A288 (1978) 27-47.

[25] Forrest, P.G.: Fatigue of metals. Pergamon Press (1962).

[26] Ostergren, W.J.; Krempl, E.: A uniaxial damage accumulation law for time-varying loading including creep-fatigue interaction. Trans. ASME, J. Pressure Vessel Technology, 101 (1979) 118-124.

[27] Goldhoff, R.M.; Woodford, D.A.: The evaluation of creep damage in a Cr-Mo-V steel, American Society for Testing and Materials, STP 515 (1972) 89-106.

[28] Burbach, J.: The physical justification of the term "state of fatigue of materials" under cyclic loading. American Society for Testing and Materials, STP 519 (1973) 185-212.

The Yield Condition of Crystalline Solids

J. LAMBERMONT

GROENE BOORD 7
6411 GE HEERLEN (HOLLAND)

Summary

A brief account of some results pertaining to the yield are presented. Accordingly to thermostatics an equilibrium state is stable when the second variation of the energy is positive definite ($\delta^2 u > 0$) and unstable when it is indefinite ($\delta^2 u$ can take negative values for one or more variables). The onset of material instability occurs when the second variation is positive -semidefinite ($\delta^2 u \geq 0$). Crystal plasticity is caused by instabilities in the dislocation structure. The situation is analyzed whereby dislocation segments bow out circularly as is the usual assumption made for the Frank-Read source ; but equally other situations as extended screw pile-ups can be considered. Applying thermodynamic formalism proves that the onset of instability is governed by calculations to be performed at constant temperature and applied stress. The resulting yield stress depends only weakly on temperature via the temperature dependence of the elastic constants. However classical thermostatics and dynamics deals only with average values of state variables. The temperature dependence of the static yield stress is explained by considering the fluctuations in the dislocation state induced by thermal agitations. This makes that a state which has a relative energy minimum and is therefore stable accordingly to classical theory may actually be unstable. Although the presented results are not new the author is not aware that the underlying assumptions made in their derivation have been proved.

Introduction

For the purpose of this paper it suffices to consider only one crystallographic slip system on which N similar dislocation segments bow out circularly by glide under an applied stress. The case is considered wherein the segments bow out between fixed pinning points, a distance l apart. In general a dislocation can bow out from the straight position in two opposite directions. The state of a dislocation belonging to a given slip system has been defined in [1] by a scalar quantity A whose absolute value is the bowed out area. For simplicity the case will be considered here that $A \geq 0$ ($\lambda = 1$) so that A denotes the average bowed out area of the dislocations (but all presented results remain valid for the general case discussed in [1]). The equations which we present apply to the macrobehaviour of a small constant mass element in a continuum ; in a polycrystalline material this particle contains still many crystallites. Under the above restrictions the fundamental energy relation and associated derivative, called the Gibbs relation, presented in [1] , reduce to

$$u = u(s, \varepsilon_{ij}, A) \tag{1}$$

$$du = T\,ds + \sigma_{ij}\,d\varepsilon_{ij} + N\eta\,dA \tag{2}$$

These equations are valid as long as no plasticity occurs, thus inside the yield surface. u and s are respectively the internal energy and entropy per unit volume of the mass element, T its absolute temperature. The macroscopic stress and strain tensor are respectively σ_{ij} and ε_{ij} ; the summation convention is used. The last terms in (2) stands for the reversible (equilibrium) work increment done on the dislocation segments per unit volume, to change their state by dA . η is the average resolved Peach-Koehler force in the Burgers vector direction acting at any of the N dislocation segments under consideration ; at equilibrium this quantity is zero [1] . By using superpositions the following expression has been derived

$$\eta = b\tau_R - b\tau \tag{3a}$$

where

$$\tau_R = \frac{DGb}{R} \quad , \quad \tau = -\sigma_{ij}\alpha_{ij} - \tilde{\tau} \qquad \text{(3b,c)}$$

R is the bowed out radius which is a function of A :

$$A = R^2 \sin^{-1}\frac{l}{2R} - \frac{l}{2}\left(R^2 - \frac{l^2}{4}\right)^{1/2} ; \; A \leq \frac{\pi}{8} l^2 \qquad (4)$$

where l is the distance between the pinning points, b the Burgers vector magnitide, G the elastic modulus of rigidity which depends generally on T or s. D is a function of Poisson's ratio and the inclination of the dislocations, its value is of unit order. The symmetric tensor

$$\alpha_{ij} = \frac{1}{2}(n_i m_j + n_j m_i) \qquad (5)$$

specifies the inclination of the slip system under consideration with respect to the macroscopic coordinate system. The unit normal to the slip system is $\underline{n}$, and $\underline{m}$ is a unit vector in the direction of the Burgers vector. The effective resolved shear stress acting at the dislocation links is τ, it is the sum of the macroscopic resolved shear stress $\sigma_{ij}\alpha_{ij}$; plus the long range stress $\tilde{\tau}$ set up by the dislocation structures, the latter changes only during plasticity. In an annealed solid $\tilde{\tau} = 0$. The minus sign has been introduced in (3a) because η =0 yields then positive values of τ. The restoring or self stress which tries to straighten the dislocations is τ_R . Let E be the self energy of a dislocation segment and L its length and e = DGb^2 its self energy per unit length, then

$$\frac{dE}{dA} = \frac{dE}{dL}\frac{dL}{dA} = \frac{DGb^2}{R} = b\,\tau_R \qquad (6)$$

At equilibrium the restoring stress balances the applied stress, a result well known in dislocation theory.

.../...

STABILITY CONSIDERATIONS ACCORDINGLY TO CLASSICAL THEORY

A sufficient condition for material stability is that the second variation of the energy is positive definite. Tisza [2] , see also Callen [3] , transformed in a general theory the quadrate form $\delta^2 u$ to the diagonal form by completion of the square. Applying this theory to (1) results in the following sufficient conditions for dislocation stability.

$$\left.\frac{\partial\eta}{\partial A}\right)_{T,\sigma_{ij}} > 0 \; ; \; \left.\frac{\partial\eta}{\partial A}\right)_{T,\varepsilon_{ij}} > 0 \; ; \; \left.\frac{\partial\eta}{\partial A}\right)_{S,\sigma_{ij}} > 0 \; ; \; \left.\frac{\partial\eta}{\partial A}\right)_{S,\varepsilon_{ij}} > 0 \qquad (7)$$

When any of these derivatives is negative the quadratic form is indefinite. The dislocation segments are unstable then i.e. act as Frank-Read sources (when they have to room to do so). When any of the quantities in (7) is zero $\delta^2 u$ can be positive-semidefinite. However as was shown by Tisza other conditions must generally be satisfied in order that it is not indefinite (takes negative values)

They are :

When $\left.\frac{\partial\eta}{\partial A}\right)_{T,\sigma_{ij}} = 0$, no restrictions (8)

when $\left.\frac{\partial\eta}{\partial A}\right)_{T,\varepsilon_{ij}} = 0$, then also $\left.\frac{\partial\eta}{\partial\varepsilon_{ij}}\right)_{T,A} = 0$ (9a,b)

when $\left.\frac{\partial\eta}{\partial A}\right)_{S,\sigma_{ij}} = 0$, then also $\left.\frac{\partial\eta}{\partial S}\right)_{A,\sigma_{ij}} = 0$ (10a,b)

when $\left.\frac{\partial\eta}{\partial A}\right)_{S,\varepsilon_{ij}} = 0$, then also $\left.\frac{\partial\eta}{\partial S}\right)_{\varepsilon_{ij},A} = 0$
and $\left.\frac{\partial\eta}{\partial\varepsilon_{ij}}\right)_{S,A} = 0$ (11a,b,c)

The question arises which of these conditions determines the onset of dislocation instability i.e. the yield condition. In fact. (9) to (11) must be rejected on physical grounds. Since the elastic modulus of rigidity which appears in (3) depends on the entropy the condition (10b) cannot be satisfied generally.

Expressing η in terms of the strains instead of the

stresses (see the equations presented in [1]) reveals that the conditions (9b) and (11c) are only satisfied when G=0. Thus the singled out yield condition is (8). Applying (8) to (3) shows that the dislocation state becomes just unstable when the dislocation segments form a semi-circle, at least when the internal stress does not vary over the position of the dislocations i.e., when $\tilde{\tau}$ does not depend on A. The corresponding critical radius and area are :

$$R_c = \frac{\ell}{2} \; ; \; A_c = \frac{\pi}{8} \ell^2 \tag{12}$$

The critical restoring self stress follows from (3b) as

$$\tau_c = \frac{2DG(T)b}{\ell} \tag{13}$$

The relation (7a) shows that for bow out less than a semi-circle the dislocations are stable.
In the phenomenological theory of plasticity the yield function and condition are introduced, these can now be defined. The yield function η_c follows from substituting (13) into (3a)

$$\eta_c \equiv b\tau_c - b\tau \; ; \; \tau = \sigma_{ij}\alpha_{ij} - \tilde{\tau} \tag{14}$$

The (phenomenological) equilibrium yield condition is

$$\eta_c = 0 \tag{15}$$

from which the equilibrium macroscopic resolved yield shear stress follows as :

$$\sigma_{ij}\alpha_{ij} = \tau_c - \tilde{\tau} \tag{16}$$

The condition (15) defines a plane in generalized stress space σ_{ij} (more generally it defines two parallel planes because dislocations can bow out in two opposite directions). Each slip system is characterized by a different α_{ij}. Thus there are as many yield functions (14) as there are slip systems in the particle under consideration. Together the yield conditions (15) define a polyhedron which is the yield surface. Inside this surface the dislocations are stable and the solid behaves in a quasi-static process elastic (reversible) but under dynamic loading ($\eta \neq 0$) the dislocations as well as heat conduction give rise to damping. This behaviour has been discussed in [1]. It is clear from the kinetic law presented there (eq. 58) that when the body is inside the yield surface and a stress is suddenly applied to it which is larger or equal than the equilibrium yield stress, it will take an incubation time before yielding occurs, i.e. before A=Ac.

TEMPERATURE DEPENDENCE OF THE YIELD STRESS.

The relations (13) to (16) show that the static yield stress depends only weakly on temperature via G. This is often contradicted by experiment. In order to incorporate the effect which fluctuations of the dislocations have on the yield stress we observe the following. The yield condition (8) is a derivative taken at constant T and σ_{ij}. This derivative follows from the Gibbs function, which is a natural function of T, σ_{ij} and A, and is defined by :

$$g = u - Ts - \sigma_{ij}\varepsilon_{ij} \tag{17}$$

Using (2) shows that

$$dg = -s\,dT - \varepsilon_{ij}\,d\sigma_{ij} + N\eta\,dA \tag{18}$$

Since

$$\frac{1}{N}\frac{\partial g}{\partial A}\Big)_{T,\sigma_{ij}} = \eta = \eta(T,\sigma_{ij},A)\,;\ \frac{1}{N}\frac{\partial^2 g}{\partial A^2}\Big)_{T,\sigma_{ij}} = \frac{\partial \eta}{\partial A}\Big)_{T,\sigma_{ij}} \tag{19}$$

dislocation stability is naturally studied in terms of g. From (18) and (3) follows that the Gibbs function, reckoned from the state for which the dislocations are straight, is

$$g = N\left\{\int_0^A \frac{DGb^2}{R(A)}\,dA - b\tau A\right\} + f(T,\sigma_{ij}) \tag{20}$$

A g-A equilibrium section is shown in Fig. 1 and 2 for the case that $\tau = \tau_c$ and $\tau < \tau_c$ respectively

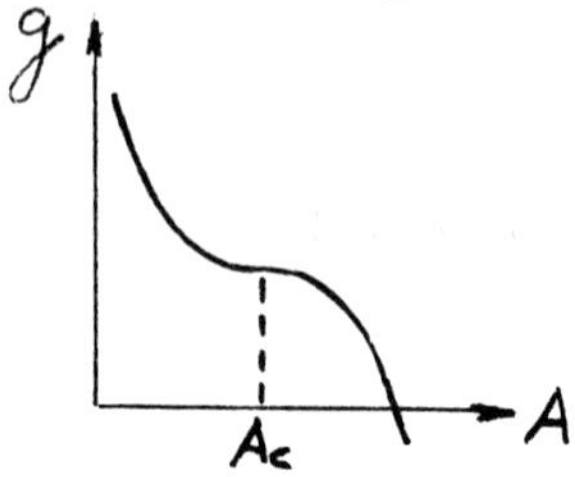

$\tau = \tau_c$

FIG. 1

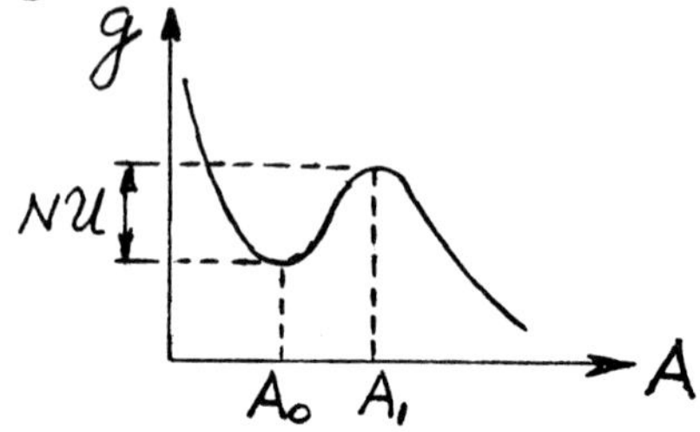

$\tau < \tau_c$

FIG. 2

At the value $A=A_c$ in Fig. 1 the Gibbs function has an inflection point with the horizontal :

$$\frac{1}{N}\frac{\partial g}{\partial A}\Big)_{T,\sigma_{ij}} = \eta = 0 \; ; \; \frac{1}{N}\frac{\partial^2 g}{\partial A^2}\Big)_{T,\sigma_{ij}} = \frac{\partial \eta}{\partial A}\Big)_{T,\sigma_{ij}} = 0 \quad \text{but} \quad \frac{1}{N}\frac{\partial^3 g}{\partial A^3}\Big)_{T,\sigma_{ij}} = \frac{\partial^2 \eta}{\partial A^2}\Big)_{T,\sigma_{ij}} \neq 0 \tag{21}$$

so that this equilibrium state is indeed unstable as the Gibbs function is not a relative minimum.

For $\tau < \tau_c$ it follows from $\eta = o$ that there are two positions of equilibrium, both are circular arcs of radius.

$$R = \frac{DGb}{l} \tag{22}$$

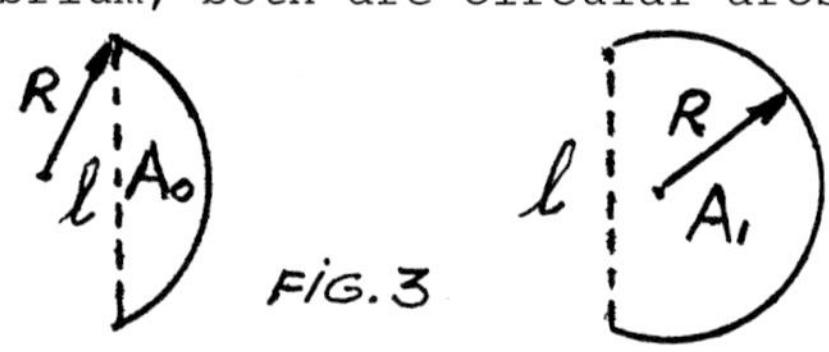

Denote the corresponding areas by A_o and A_1. The state A_o is stable, $\frac{\partial \eta}{\partial A}\Big)_{T,\sigma_{ij}} > 0$ while that corresponding with A_1 is unstable $\frac{\partial \eta}{\partial A}\Big)_{T,\sigma_{ij}} < 0$. A_o and A_1 are functions of τ and depend only weakly on T via G. In the region $0 \leq A < A_1$, fig. 2, the dislocations are stable while for $A \geq A_1$ they are unstable. When due to the thermal agitation of the atoms a dislocation fluctuation occurs of such strength that it moves from the stable equilibrium position Ao to the saddle point A_1, it will expand spontaneously. A dislocation loop is then created giving rise to plasticity. The yield stress will then be lower than predicted by eq. (21), i.e. by classical theory.

.../...

From Einstein's fluctuation theory [3],[4] it follows that when the macroscopic stress and temperature are kept constant the probability of finding a dislocation in the range A, A + dA is

$$W(A) \propto \exp - \frac{g/N}{kT} \tag{22}$$

where k is Boltzamann's constant.

Hence
$$\frac{W(A_o)}{W(A)} = \exp - \mathcal{U}/kT \,, \tag{23}$$

where

$$\mathcal{U} = \frac{1}{N} \Delta g \Big)_{T,\sigma_{ij}} = \int_{A_o}^{A_1} \eta(T,\sigma_{ij},A)\, dA \tag{23a}$$

is the activation energy per dislocation segment. It follows from (18) that it is the reversible work done at a link to bow it out from A_o to A_1, when T and σ_{ij} are kept equal to the values of the considered stable equilibrium state.

In the approximation of circular bow out, dislocations vibrate with an average circular frequency of [5].

$$\nu_o \simeq \nu_D \frac{b}{l} \tag{24}$$

where ν_D is the Debey frequency ($\nu_D \simeq 10^{13}/sec.$). The average period of vibration is $t_o = 1/2\pi\nu_o$. Let t_1 be the mean waiting time required to cross the energy barrier, thus the time it takes for a fluctuation of energy $\mathcal{U}$ to occur. Then,

$$\frac{W(A_o)}{W(A_1)} = \frac{t_1}{t_o} \tag{25}$$

and thus

$$t_1 \simeq \frac{l}{2\pi b \nu_D} \exp \frac{\mathcal{U}}{kT} \tag{26}$$

or

$$\mathcal{U} = mkT \text{ where } m \simeq \ln \frac{2\pi b \nu_D t_1}{l} \tag{27}$$

Taking typically l=10000 b it follows from (26) that at room temperature (T=300°K, kT≈0.024 ev) the mean waiting time for a fluctuation with an energy of 1ev to occur is about 10^6 sec. while for 1.25ev it takes about 10^{10} sec. Thus fluctuations with an energy larger than 1ev hardly occur in nature at temperatures less than room temperature. We are interested in the yield stress of quasi-static or time independent plasticity. Thus in long waiting times. For $t_1 = 10^4$ sec. $\simeq$ 3 hours, $m \simeq 35$, for $t_1 = 10^6$ sec. $\simeq$ 11 days, $m \simeq 40$ and for $t_1 = 10^8$ sec $\simeq$ 3 years, $m \simeq 44$. Whence selecting $m = 40$ in (27) corresponds with long waiting times.

Nabarro [6] has calculated with (27) the temperature dependent critical restoring stress τ_{c_T} for fixed pinning points. When its value is closed to τ_c the results is

$$\tau_{c_T} = \tau_c \left[1 - \left(\tau_c \frac{3bm\,kT}{8\sqrt{2}\,(DGb^2)^2} \right)^{2/3} \right] \tag{28}$$

The corresponding temperature dependent critical radius R_{c_T} follows from (3) as

$$R_{c_T} = \frac{DGb}{\tau_{c_T}} \tag{29}$$

The corresponding area $A_{c_T} < A_c$ follows from (4). To incorporate the effect of stress assisted thermal fluctuations the yield function (14) must be modified to read

$$\eta_c = b\,\tau_{c_T} - b\,\tau \; ; \quad \tau = -\sigma_{ij}\alpha_{ij} - \tilde{\tau} \tag{30}$$

The static yield condition is $\eta_c = 0$

Substituting $\tau_c = 1\ kg/mm^2$ and the typical values $G = 4000\ kg/mm^2$, $b \simeq 10^{-7}$ mm, $D \simeq 1$ shows that for $m = 40$ the lowering of the yield stress is about 0,5% at room temperature. However when the distance between the pinning points becomes smaller, as is the usual situation during plasticity, and thus by (13) τ_c becomes larger the effect increases. For $l = 1000b$ leading to the large value $\tau_c \simeq 10\ kg/mm^2$ the lowering is about 2% at room temperature. For the unrealistic value of $l = 100b$ the lowering is 12%. This shows that thermal fluctuations are only effective for objects comprising a linear dimension less than some thousand atom distances, as anchoring points. For example, edge multipole bundles contain generally points defect clusters which may act as anchoring points from which dislocations bow out. To calculate the breakage stress τ_{c_T} requires a rather detailed knowledge of its constitution but one expects generally that stress induced thermal activation leads here to a strong temperature dependence of the yield stress.

References

1 Lambermont, J. : Dislocations as internal variables in the Thermodynamic theory of anelasticity, To be published in: Continuum models of discrete system, 3rd Int. Symp. ed. E. Kroner. Freudenstadt (1979).

2 Tisza, L. : On the general theory of phase transitions. Chapter I in phase transformations in solids. eds. Smoluchowski, Mayer, Weyl, Wiley (1960).

3 Callen, H.B. : Thermodynamics. Wiley (1960).

4 Reif, F. : Fund amentals of statistical and thermal physics. Mc Graw-Hill(1965)

5 Friedel, J. : Dislocations. Pergamon Press (1964)

6 Nabarro, F.R.N.: The mathematical theory of stationary dislocations. In advances in physics. Taylor and Francis Limited. London (1952) 1, 269.

=-=-=-=-=-=-=

Tensorial Nature of Damage Measuring Internal Variables

F. A. LECKIE

Department of Theoretical and Applied Mechanics and
Department of Mechanical and Industrial Engineering
University of Illinois at Urbana-Champaign
Urbana, IL 61801

E. T. ONAT

Department of Engineering and Applied Science
Yale University
New Haven, CN 06520

Summary

A theoretical framework is developed which can be used to describe the constitutive equations of creeping materials which also suffer internal damage. Within this framework, specialized constitutive equations are derived for a nickel alloy which has been subjected to non-proportional histories of stress.

1. Introduction

Metals when subjected to elevated temperatures (above 0.3 T_m where T_m is the melting temperature of the basic constituent) and stress undergo time-dependent deformation called creep. At stress levels frequently encountered in engineering components, the nucleation and growth of grain boundary defects produce an increase in the creep rate and eventual rupture. There is a large body of metallurgical and phenomenological work, starting with Kachanov [1], to present research where electron microscopy plays a central role (cf. [2]) devoted to the study of the nature of internal damage (i.e., the presence of cavities and nucleation sites on grain boundaries) and its effects on mechanical behavior.

This paper is concerned with the phenomenological description of mechanical behavior in the presence of internal damage. This subject was first studied [1] where a scalar was used to represent internal damage. But recent work suggests, as we demonstrate in Section 2, that the distribution of voids within a deforming element is anisotropic and that this anisotropy could play an important role in behavior under non-proportional loading. Hence the need to reconsider the problem of representation of damage. Section 3 is devoted to a study of this problem. It is shown that the distribution of void volume among grain boundaries can be represented by a sequence of even rank-irreducible tensors. This result supports earlier work on the general theory of state variable representation of mechanical behavior [3]. In Section 4 we introduce simplifications and give a fairly complete representation of mechanical behavior based on a second rank-symmetric tensor that measures (approximately) the distribution of void volume within a deforming element.

2. Tertiary Creep of Nimonic

A high temperature creep resistant nickel base alloy, Nimonic 80A, is selected as the example material to be considered in this paper. The reason for this is that this material has been studied extensively both from microscopic and macroscopic points of view [4,5,6]. Detailed quantitative observations with a 1 MV microscope by Dyson, et al. [4], showed that preferential cavity growth occurred during creep: cavities on grain boundaries which were orthogonal to the applied tension σ_{max} grew relatively quickly while cavities on boundary parallel to applied tension either remained constant in size or diminished. It is also shown in this work that the availability of sites for void or cavity growth is of central importance and that these sites can be created by plastic deformation at room temperature by grain interior slip.

Dyson and McLean [5] proposed equations for the rate of void nucleation and void nucleation and void growth based on their experiments conducted under constant uniaxial stress or pure shear.

In Fig. 1, the effective strain-time curves are presented for two creep tests performed at 750°C: (a) uniaxial tension at effective stress $\bar{\sigma}$ = 309 MPa [5] and (b) torsion at $\bar{\sigma}$ = 297 MPa [6]. The same material and same procedures for specimen preparation were used in both test.

The effective stress $\bar{\sigma}$ (=$[3/2 \text{ tr } s^2]^{1/2}$ where $\underset{\sim}{s}$ denotes the deviatoric stress tensor) is nearly the same in these tests but σ_{max} is approximately $3^{1/2}$ times greater in test (a). Observe now from

Fig. 1 that in test (a), the rate of strain is higher than in test (b) and rupture occurs much earlier and at a much smaller strain. Combining the above remarks, we conclude that tests (a) and (b) support the above mentioned microscopic observations concerning the importance of σ_{max} in the development of internal damage in Nimonic.

On the other hand, it is known [7] that for most materials, including Nimonic, $\dot{\underset{\sim}{\varepsilon}}_c$, the time rate of creep strain depends (at a given time and to within an acceptable error) only on the deviatoric part s of the stress and in the following manner:

$$\dot{\underset{\sim}{\varepsilon}}_c = F \underset{\sim}{s} \qquad (2.1)$$

where F is a scalar valued function of $\underset{\sim}{s}$ and of internal damage that is present within the material in the form of intergranular cavities.

We see from Eq. (2.1) that the effect of the average normal stress (i.e., 1/3 tr $\underset{\sim}{\sigma}$) on the creep behavior will be felt only through the dependence of F on the internal damage, the development of internal damage being controlled by σ_{max} and its direction.

Hayhurst, Trampczynski, and Leckie [6] conducted reverse torsion tests at 750°C (Fig. 2). In a reverse torsion, test the direction of σ_{max} rotates 90° upon the reversal of stress. The ensuing void distribution could, therefore, be significantly different from that encountered in simple torsion tests. Hence the importance of such tests in the study of the dependence of F on internal damage.

Curve I in Fig. 2 is obtained from a simple torsion test at $\bar{\sigma}$ = 297 MPa and it is included for purposes of comparison. Curve II represents the results of a reverse torsion experiment at $\bar{\sigma}$ = 297 MPa where the shear stress was altered in sign at t = 200 hours. We see that upon reversal of stress, a small transient creep of short duration takes place followed by tertiary creep in reverse direction. Moreover, it is seen that

$$\dot{\varepsilon}_{II}(t) \simeq -\dot{\varepsilon}_I(t);\ 290 > t > 210 \text{ hrs.} \qquad (2.2)$$

This remarkable circumstance implies that F of Eq. (2.1) has the same momentary value in tests I and II on the above time interval.

However, it is seen that the specimen subjected to reversal of stress has a rupture life that is considerably longer than the specimen of test I.

3. Representation of Mechanical Behavior in the Presence of Internal Damage

For the sake of simplicity, we consider here only the case of small isothermal deformations and rotations, although the experiments discussed above go beyond small deformations and some of them contain episodes at room temperature. We assume that the material is simple, initially isotropic, and non-aging under zero stress.

The state variable representation of behavior which we shall employ here is based on the notion that the internal state and orientation of a deforming material element can be described, to within any desirable accuracy, by a number of parameters. These may be thought of as the n-components of a vector S in R^n. In the present context, it is useful to identify and to order these parameters as follows: $S = (\underset{\sim}{\sigma}, \underset{\sim}{q}, \underset{\sim}{r})$ where $\underset{\sim}{\sigma}$ denotes the stress tensor that the element carries. $\underset{\sim}{\sigma}$ may be thought of as representing, through Hooke's law, the strain of the underlying average lattice. $\underset{\sim}{q}$ stands for those parameters that measure the relevant aspects of the distribution of dislocations within the element. Finally, $\underset{\sim}{r}$ constitutes a set of parameters that measure intergranular events of interest such as the size and distribution of voids and the number and distribution of sites for void growth. It should be noted that these parameters measure not only the magnitude of quantities of interest but also their "directions." The stress tensor $\underset{\sim}{\sigma}$ provides a clear example of this by its principal values and directions.

To amplify this last point and to explore the implications of the assumed isotropy, consider two experiments in which the following strain histories are applied to identical virgin specimens:

$$\underset{\sim}{\varepsilon}_I = \underset{\sim}{\varepsilon}(t), \ \underset{\sim}{\varepsilon}_{II}(t) = \underset{\sim}{Q}\,\underset{\sim}{\varepsilon}(t)\,\underset{\sim}{Q}^T, \qquad \underset{\sim}{Q} \in 0^+(3), \ t \in [0,t_1] \tag{3.1}$$

where $\underset{\sim}{\varepsilon}(t)$ stands for the components $\varepsilon_{ij}(t)$ of the infinitesimal strain tensor in a fixed rectangular frame at time t. Moreover, in (3.1), $\underset{\sim}{Q}\,\underset{\sim}{\varepsilon}\,\underset{\sim}{Q}^T$ denotes the components $\Sigma^3_{p,q}\, Q_{ip} Q_{jp}\, \varepsilon_{pq}(t)$ where Q_{ij} is a fixed proper orthogonal transformation on R^3.

Consider now the coordinate transformation $\underset{\sim}{x}' = \underset{\sim}{Q}^T \underset{\sim}{x}$ and observe from (3.1) that the components of the strain tensor $\underset{\sim}{\varepsilon}_{II}$ in the new (primed) coordinate system will be identical with those of $\underset{\sim}{\varepsilon}_I$ in the original frame:

$$\underset{\sim}{\varepsilon}'_{II} = \underset{\sim}{\varepsilon}(t). \tag{3.2}$$

The initial isotropy of the material together with (3.1) and (3.2) imply that from the point of view of an observer working in the primed coordinate frame, test II will be identical with test I (as observed in the original frame) and will produce identical results (Fig. 3).

It should be clear from these remarks that during these tests, the two specimens will carry, in the original frame, the stresses $\underset{\sim}{\sigma}$ and $\underset{\sim}{Q}\,\underset{\sim}{\sigma}\,\underset{\sim}{Q}^T$, respectively, and that their internal states will differ only in orientation. (For instance, if severe void formation took place on certain grain boundaries in test I (parallel lines in Fig. 3), then it will be found that in test II, severe formation will occur on another set of grain boundaries, the two sets differing from each other by the rotation $\underset{\sim}{Q}$ in orientation. It should be emphasized, however, that because of this difference in orientation, the two elements would, in general, behave differently beyond time t_1 under identical strain histories.

We let S_I and S_{II} denote the state and orientation of the two elements at time t_1 and observe, in view of the very last remark, that $S_I \neq S_{II}$, in general. But S_{II} is closely related to S_I. We can create an element in S_{II} by simply rotating an element in S_I by the amount $\underset{\sim}{Q}$ (cf. Fig. 3). Thus, $S_{II} = P(\underset{\sim}{Q}, S_I)$ and $P\colon 0^+(3) \times R^n \to R^n$. This can be written as

$$S_{II} = P_{\underset{\sim}{Q}}\, S_I \text{ and } P_{\underset{\sim}{Q}}\colon R^n \to R^n \tag{3.3}$$

and $P_{\underset{\sim}{Q}}$ can be regarded as a transformation on R^n that is induced by the rigid body rotation $\underset{\sim}{Q} \in 0^+(3)$ of the element.

We know that Nimonic develops anisotropy during tertiary creep so that, in general, $P_{\underset{\sim}{Q}}\, S(t) \neq S(t)$. Thus, in order to specify the degree of anisotropy, we must be able to specify the action of $P_{\underset{\sim}{Q}}$ in the state and orientation space R^n.

It is shown in [3] that the action of $P_{\underset{\sim}{Q}}$ on R^n can be conveniently described by means of tensorial state variables and as follows: First S is taken to be composed of a number of (irreducible) tensors of even rank

$$S = (\underset{\sim}{\sigma};\ \underset{\sim}{q}_1,\ldots,\underset{\sim}{q}_n;\ \underset{\sim}{r}_1,\ldots,\underset{\sim}{r}_m) \tag{3.4}$$

and then $P_{\underset{\sim}{Q}}\, S$ is defined by means of ordinary tensor transformations:

$$P_{\underset{\sim}{Q}}\, \underset{\sim}{S} = (P_{\underset{\sim}{Q}}\, \underset{\sim}{\sigma};\ P_{\underset{\sim}{Q}}\, \underset{\sim}{q}_1,\ldots,P_{\underset{\sim}{Q}}\, \underset{\sim}{q}_n;\ P_{\underset{\sim}{Q}}\, \underset{\sim}{r}_1,\ldots,P_{\underset{\sim}{Q}}\, \underset{\sim}{r}_m) \tag{3.5}$$

where say $P_{\underset{\sim}{Q}}\, \underset{\sim}{r}_i$ has the meaning of the tensor transformation appropriate to the rank of r_i. For instance, $\underset{\sim}{\sigma}$ is a second rank tensor and $P_{\underset{\sim}{Q}}\, \underset{\sim}{\sigma} = (\Sigma_{p,q}\, Q_{ip}\, Q_{jq}\, \sigma_{pq}) = \underset{\sim}{Q}\, \underset{\sim}{\sigma}\, \underset{\sim}{Q}^T$.

The next major step in the state variable representation of mechanical behavior is the establishment of the laws of growth (or evolution) for the state and orientation S. In the present work, we treat this step only for the case of <u>slowly varying deformations and stress</u>.

We start by assuming, with some justification, that $\underset{\sim}{q}^{\cdot}$, the time rate of change of parameters that govern the distribution of dislocation throughout the body, depends strongly on the present values of the stress $\underset{\sim}{\sigma}$ and $\underset{\sim}{q}$ but weakly on $\underset{\sim}{r}$, intergranular happenings. We further assume that under constant stress, $\underset{\sim}{q}(t)$ attains it saturation value within a time that is small compared with the time to failure. Under these assumptions, one can make the assertion that in the presence of slowly varying stresses

$$\underset{\sim}{q}(t) = F[\underset{\sim}{\sigma}(t)]. \tag{3.6}$$

(It will be remembered that in test II described in Fig. 2, the stress is changed abruptly at 200 hours. This causes $\underset{\sim}{q}$ to change from one equilibrium value to another one over the course of time and thereby creates transient creep visible in Fig. 2).

For slowly varying loads and using the assumptions that led to (3.6), the growth laws for $\underset{\sim}{\sigma}$ and $\underset{\sim}{r}$ can be written as follows:

$$d\underset{\sim}{\sigma}/dt = \underset{\sim}{f}[\underset{\sim}{\sigma}(t), \underset{\sim}{r}(t)] + \underset{\sim}{E}(d\underset{\sim}{\varepsilon}/dt) \qquad \underset{\sim}{f}: T_2 \times R^{\ell} \to T_2$$

$$d\underset{\sim}{r}/dt = \underset{\sim}{g}[\underset{\sim}{\sigma}(t), \underset{\sim}{r}(t)] \qquad \underset{\sim}{g}: T_2 \times R^{\ell} \to R^{\ell} \tag{3.7}$$

where $\underset{\sim}{E}$ is the fourth rank tensor of elastic coefficients ($E_{ijk\ell} = 2\mu\, \delta_{ij}\, \delta_{k\ell} + \lambda\, \delta_{ik}\, \delta_{j\ell}$) and T_2 denotes the space of symmetric second rank tensors. Here ℓ is the number of parameters that comprise $\underset{\sim}{r}$. Note that

$$\underset{\sim}{r} = (\underset{\sim}{r}_1, \ldots, \underset{\sim}{r}_m) \tag{3.8}$$

and $m \ll \ell$ when $\underset{\sim}{r}_i$'s are not all scalars.

It is customary to append to (3.7) a rupture criterion in the form of an inequality

$$R(\underset{\sim}{\sigma}, \underset{\sim}{r}) \leq 0; \qquad R: T_2 \times R^{\ell} \to R \tag{3.9}$$

and to assert that rupture will occur when R reaches the value 0.

In view of our remarks concerning the tensorial nature of state variables and the role of initial isotropy, the functions f, g, and R are subject to the invariance requirements listed below:

$$\underset{\sim}{f}(P_{\underset{\sim}{Q}}\, \underset{\sim}{\sigma}, P_{\underset{\sim}{Q}}\, \underset{\sim}{r}) = P_{\underset{\sim}{Q}}\, \underset{\sim}{f}(\underset{\sim}{\sigma}, \underset{\sim}{r}) = \underset{\sim}{Q}\, \underset{\sim}{f}(\underset{\sim}{\sigma}, \underset{\sim}{r})\, \underset{\sim}{Q}^T;$$

$$\underset{\sim}{g}(P_{\underset{\sim}{Q}}\, \underset{\sim}{\sigma}, P_{\underset{\sim}{Q}}\, \underset{\sim}{r}) = P_{\underset{\sim}{Q}}\, \underset{\sim}{g}(\underset{\sim}{\sigma}, \underset{\sim}{r}); \tag{3.10}$$

$$R(P_{\underset{\sim}{Q}}\, \underset{\sim}{\sigma}, P_{\underset{\sim}{Q}}\, \underset{\sim}{r}) = R(\underset{\sim}{\sigma}, \underset{\sim}{r}). \tag{3.11}$$

We recall from (3.5) that the action of $P_{\underset{\sim}{Q}}$ in R^{ℓ} is defined as follows:

$$P_{\underset{\sim}{Q}}\, \underset{\sim}{r} = (P_{\underset{\sim}{Q}}\, \underset{\sim}{r}_1, \ldots, P_{\underset{\sim}{Q}}\, \underset{\sim}{r}_m). \tag{3.12}$$

The mathematical structure, Eqs. (3.7) through (3.12), constitutes only a starting point in the task of representing the mechanical behavior of a given material such as Nimonic. We must now undertake the difficult job of choosing the number and the tensorial nature of parameters $(\underset{\sim}{q}_1, \ldots, \underset{\sim}{q}_m)$ that define the relevant aspects of intergranular events. Here our guide will be experiments of both microscopic and phenomenological nature. It is known from experiments that intergranular voids are of central importance, hence the need to describe the relevant aspects of void distributions within the material element. (Here we assume that our element is big enough to contain a very large number of grains and that its "average" deformation as measured by the strain tensor $\underset{\sim}{\varepsilon}$ is homogenous.) Clearly the total volume of voids within an element of unit volume is of basic significance. But so is the distribution of this volume among the grain boundaries. (If one wants to introduce more details, then one could worry about the shape and size of the individual voids.) To be more specific about these points, consider a sphere of unit radius and let the position vector of each point of this sphere represent a space direction $\underset{\sim}{n}$. An infinitesimal area dA on this sphere about point $\underset{\sim}{n}$ represents a bundle of directions about $\underset{\sim}{n}$. Now consider a spherical material element and focus the attention on the grain boundaries that are orthogonal to the directions present within the bundle.

The total volume of voids that are found in these grain boundaries will be given by

$$\hat{V}(\underset{\sim}{n})\ dA(\underset{\sim}{n}) \tag{3.13}$$

where $\hat{V}$ is density of the distribution of void volume and $\hat{V}$ will, in general, depend on the direction $\underset{\sim}{n}$. It is obvious that $\hat{V}$ has the property

$$\hat{V}(\underset{\sim}{n}) = \hat{V}(-\underset{\sim}{n}). \tag{3.14}$$

One can try to represent $\hat{V}$: $S_2 \to R$ where S_2 is the unit sphere in terms of its moments:

$$\hat{V}_o = \int_{S_2} \hat{V}(\underset{\sim}{n})\ dA(\underset{\sim}{n}),\ \hat{V}_i = \int_{S_2} \hat{V}(\underset{\sim}{n})\ n_i\ dA(\underset{\sim}{n}),$$

$$\hat{V}_{ij} = \int_{S_2} \hat{V}(n)\ n_i\ n_j\ dA(\underset{\sim}{n}).. \tag{3.15}$$

where n_i denotes the components of $\underset{\sim}{n}$ in a fixed rectangular frame $(i,j,\ldots = 1,2,3)$.

Observe first that these quantities behave as tensors under rigid rotations of the element. Moreover, it is readily seen from (3.14) that the odd rank tensors in (3.15) vanish:

$$\hat{V}_i = \hat{V}_{ijk} = \ldots = 0 \tag{3.16}$$

and the even rank tensors are completely symmetric:

$$\hat{V}_o, \hat{V}_{ij} = \hat{V}_{ji}, \hat{V}_{ijk\ell} = \hat{V}_{jik\ell} = \hat{V}_{jki\ell} = \text{etc.},\ldots \tag{3.17}$$

Note that $\hat{V}_0$ measures the total volume of voids per unit volume of material. Moreover, we have by definition

$$\Sigma_k \hat{V}_{kk} = \hat{V}_o, \Sigma_k \hat{V}_{ijkk} = \hat{V}_{ij},\ldots \tag{3.18}$$

Thus, the sequence of tensors in (3.15) can be replaced by traceless even rank tensors that are independent from each other. It is worth noting that after this final step, we end up with a sequence of irreducible even rank tensors as state variables as expected from the general theory (cf. (3.1)).

On the other hand, experiments also suggest (cf. Section 2) that the presence (or absence) of sites for void growth are of basic importance. Therefore, we are obliged to introduce another distribution:

$$\hat{N}: S_2 \rightarrow R \tag{3.19}$$

that measures the number of sites available for void growth on grain boundaries perpendicular to a given direction n. This would lead, following steps similar to those found in (3.13) through (3.17), to a hierarchy of tensors:

$$\hat{N}_o, \hat{N}_{ij}, \hat{N}_{ijk},\ldots \tag{3.20}$$

Thus, we arrive at the conclusion that intergranular events can be described (at least partially) by the sequence of physically significant tensors given in (3.17) and (3.20):

$$\mathfrak{X} = (\hat{V}_o, \hat{V}_{ij}, \hat{V}_{ijk\ell},\ldots; \hat{N}_o, \hat{N}_{ij}, \hat{N}_{ikj\ell},\ldots) \tag{3.21}$$

This closes the cycle of ideas that started with Eq. (3.6). In order to make further progress, we must be more specific about (3.7) and (3.9) and must introduce simplifications so as to arrive at a workable representation.

4. Simplifications

This section is intended as an example of simplifying steps that one might take to arrive at a manageable (and therefore approximate) representation for a given material such as Nimonic.

The first step is to approximate (3.21) by a few tensors. Here we propose to use <u>two symmetric second rank tensors</u>: V_{ij} and N_{ij}:

$$\underset{\sim}{r} = (\underset{\sim}{V},\underset{\sim}{N}). \tag{4.1}$$

We let tr $\underset{\sim}{V} = V_{11} + V_{22} + V_{33}$ to represent the total volume $\hat{V}_o$ of voids within an element of unit volume. Moreover, we think of

$$3/4\pi \; \Sigma_{i,j} \; V_{ij} \; n_i \; n_j \tag{4.2}$$

as an approximate representation of the void distribution function $\hat{V}(\underset{\sim}{n})$ (cf. (3.13)).

Thus, accordingly,

$$\hat{V}_o = \text{tr } \underset{\sim}{V}, \quad \hat{V}(\underset{\sim}{n}) \simeq 3/4\pi \; \Sigma_{i,j} \; V_{ij} \; n_i \; n_j \tag{4.3}$$

We recall from (3.15) that $\hat{V}_o = \int_S \hat{V}(\underset{\sim}{n}) \, dA(\underset{\sim}{n})$. It is easily seen that (4.3) satisfies this consistency condition. (we assign similar meanings to tr $\underset{\sim}{N}$ and $3/4\;\pi\Sigma \; N_{ij} \; n_i \; n_j$; thus, tr $\underset{\sim}{N}$ represents the total number of sites for void growth per unit volume of material, etc.).

It should be observed that the choice of V_{ij} as a tensorial state variable and the interpretation given to it by (4.3) allows us to introduce anisotropy in the distribution of voids within the material element. One might think that one could achieve this more simply by state variables that are vectors or antisymmetric second rank tensors. But as (3.16) and (3.17) assert, these tensors are not appropriate for the representation of the distribution of voids. On the other hand, the fact that we do not use fourth and higher rank tensors here is a source of error and approximation.

Next we consider the growth laws (3.7). The essence of the first of (3.7) is $\underset{\sim}{f}(\underset{\sim}{\sigma},\underset{\sim}{r})$ or equivalently $-\underset{\sim}{E}^{-1} \underset{\sim}{f}$, the latter representing the easily observable creep strain rate $\dot{\varepsilon}_c = -\underset{\sim}{E}^{-1} \underset{\sim}{f}$. In view of our previous comments on creep strain rate (cf. (2.1)) and (4.3), we expect $\dot{\underset{\sim}{\varepsilon}}_c = f(\underset{\sim}{s},\underset{\sim}{V}) \, \underset{\sim}{s}$ where as before $\underset{\sim}{s} = \underset{\sim}{\sigma} - (1/3) \, \underset{\sim}{I} \text{ tr } \underset{\sim}{\sigma}$ is the deviatoric stress tensor and f is a scalar valued function of the joint invariants of $\underset{\sim}{s}$ and $\underset{\sim}{V}$. (Note the absence of $\underset{\sim}{N}$.) Here we propose to use the simpler form

$$\dot{\underset{\sim}{\varepsilon}}_c = \underset{\sim}{f}(J_2, \operatorname{tr} \underset{\sim}{V}) \underset{\sim}{s} \tag{4.4}$$

where $J_2 = 1/2 \operatorname{tr} \underset{\sim}{s}^2$. This form implies that in the reverse torsion experiments, the strain rate would just change sign upon the reversal of the stress. We note that the observation (2.2) concerning the strain rates in Fig. 2 supports this conclusion.

The representation is completed by writing the growth laws for $\underset{\sim}{V}$ and $\underset{\sim}{N}$:

$$d\underset{\sim}{V}/dt = \underset{\sim}{g}_V(\underset{\sim}{\sigma},\underset{\sim}{V},\underset{\sim}{N}), \quad d\underset{\sim}{N}/dt = \underset{\sim}{g}_N(\underset{\sim}{\sigma},\underset{\sim}{V},\underset{\sim}{N}) \tag{4.5}$$

and the rupture law $R(\underset{\sim}{\sigma},\underset{\sim}{V}) \leq 0$. These are subject to invariance requirements (3.10). It is our belief that the functions $\underset{\sim}{g}_V$, $\underset{\sim}{g}_N$, and $\underset{\sim}{R}$ can be determined with the help of well-planned phenomenological and microscopic observations. We shall pursue these thoughts in another paper. Here we limit ourselves to a consideration of simplest possible growth and rupture laws suggested by the experimental evidence discussed in Section 2.

We assume that for virgin Nimonic (i.e., Nimonic that has not been subjected to large cold deformations), the sites for potential void growth are uniformly distributed throughout the material and their number does not vary appreciably during loading. We further assume that

$$d\underset{\sim}{V}/dt = \underset{\sim}{g}_V(\underset{\sim}{\sigma}). \tag{4.6}$$

In view of the invariance requirements discussed earlier, it follows then that the principal directions of V must coincide with those of $\underset{\sim}{\sigma}$. Denoting the principal values by V_i and σ_i, (4.6) can be written as

$$\dot{V}_i = g_i(\sigma_j) \qquad i,j = 1,2,3 \tag{4.7}$$

where g_i are subject to invariance requirements $g_1(\sigma_1,\sigma_2,\sigma_3) = g_2(\sigma_2,\sigma_1,\sigma_3)$, etc. Preferential growth of voids in the direction of maximum tensile stress suggests that we choose the following form for (4.7), [8].

$$V_i^{\cdot} = g(\sigma_i/\sigma_o) \tag{4.8}$$

with $g(x) > 0$ when $x > 0$ and $g(x) = 0$ when $x \leq 0$. This assures us that there is no void growth in the direction of compressive stresses. If in addition one takes $g(x) \simeq x^n$, $n >> 1$, then the void growth would essentially be controlled by the direction and magnitude of σ_{max} and σ_o would play the role of a threshold stress for void growth.

Finally we offer a rupture criterion of the form

$$R(\sigma_{max}, \operatorname{tr} \underset{\sim}{V}, \Sigma_{i,j} V_{ij} n_i^* n_j^*) = 0 \tag{4.9}$$

where $\underset{\sim}{n}^*$ is the direction of the maximum principal stress and hence $\sigma_{max} = \Sigma_{i,j} \sigma_{ij} n_i^* n_j^*$. According to this criterion, rupture occurs when the maximum stress, the total void volume and the void density on the grain boundaries orthogonal to the maximum stress reach a critical combination. The criterion is suggested by the reverse torsion experiments as we presently shall see.

We have arrived at a complete and simple description of mechanical behavior in terms of Eqs. (4.4), (4.8), and (4.9). We need not be much more specific about these equations in order to predict the broad aspects of behavior. In order to clarify this point, we consider the predictions of (4.4), (4.8), and (4.9) for the reverse torsion test discussed in Section 2 (cf. Fig. 2). We first observe that the direction of principal stress remain fixed during these tests. Moreover, $\sigma_1 + \sigma_2 = 0$, $\sigma_3 = 0$ and $\sigma_1 = \tau_o$ in test I and $\sigma_1 = \tau_o$ on (0,200 hours) and $\sigma_1 = -\tau_o$ beyond 200 hours in test II. The evolution of the principal values V_i of the tensor $\underset{\sim}{V}$ follows from (4.8). We have, of course, $V_3 = 0$ and the evolution of V_1 and V_2 can be discussed conveniently in terms of a geometric picture (Fig. 4). In this figure, OAR_1, represents the image of test I where A corresponds to 200 hours at R_1 is the rupture point. On the other hand, OAR_2 is the image of test II. Points B_1 and B_2 correspond to the same time t in tests I and II. We see from (4.8) that $AB_1 = AB_2$ and this implies that $\operatorname{tr} \underset{\sim}{V} = V_1 + V_2$ has the same value in both tests at a given time. This observation, when coupled with (4.4), yields $\varepsilon_I^{\cdot}(t) = -\varepsilon_{II}^{\cdot}(t)$ when $t > 200$ hours which is supported by experiments (cf. Fig. 2 and Eq. (4.1)).

The time to rupture is longer in test II, hence point R_2 must lie beyond the 45° line passing through R_1. This implies that a rupture criterion such as $R(\sigma_{max}, tr\ V) \leq 0$ would not be appropriate. It is this observation that led us to the rupture criterion (4.9).

Acknowledgements

This work was started when E. T. Onat was Visiting Professor at the Department of Theoretical and Applied Mechanics at the University of Illinois at Urbana-Champaign, Urbana, Ill., during May 1979.

F. A. Leckie acknowledges the support of the National Science Foundation under Grant NSF ENG 78 23549.

References

1. Kachanov, L. M. (in Russian), Izv ARad. Nauk SSSR otd. Tekh. Nauk No. 8 (1958).

2. Leckie, F. A., and D. R. Hayhurst, Acta Metallurgica, 25, 1959 (1977).

3. Geary, J. A., and E. T. Onat, Oak Ridge National Laboratory, ORNL-TN-4525, June 1974.

4. Dyson, B. F., M. S. Loveday, and M. J. Rodgers, Proc., Roy. Soc., London A, 349, 245 (1976).

5. Dyson, B. F., and D. McLean, Metal Sci., 2, 37 (1977).

6. Hayhurst, D. R., W. A. Trampczynski, and F. A. Leckie, "Creep Rupture under Non-Proportional Loading" (to appear in Acta Metallurgica).

7. Leckie, F. A., and D. R. Hayhurst, Proc., Roy. Soc., London A, 340 323 (1974).

8. Hayhurst, D. R. and B. Storåkers, Proc., Roy. Soc., London A, 349 369 (1976).

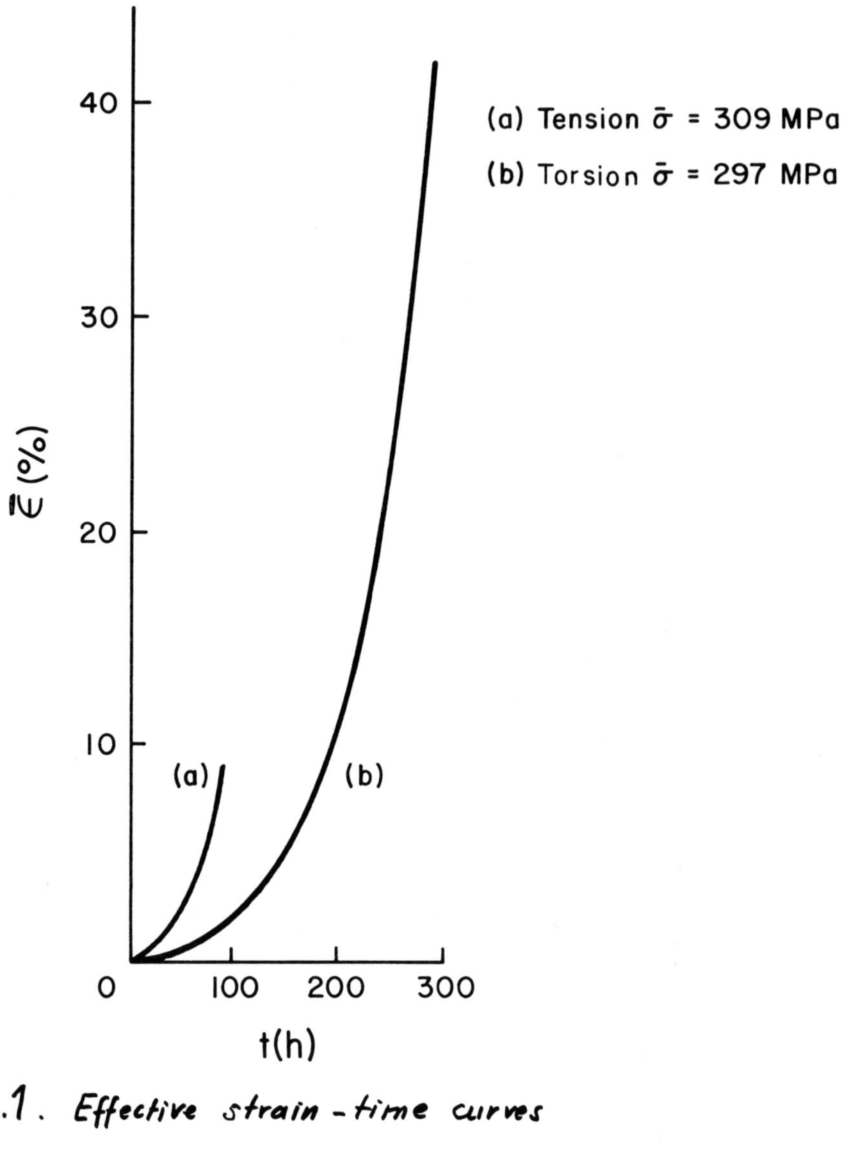

Fig. 1. *Effective strain - time curves*

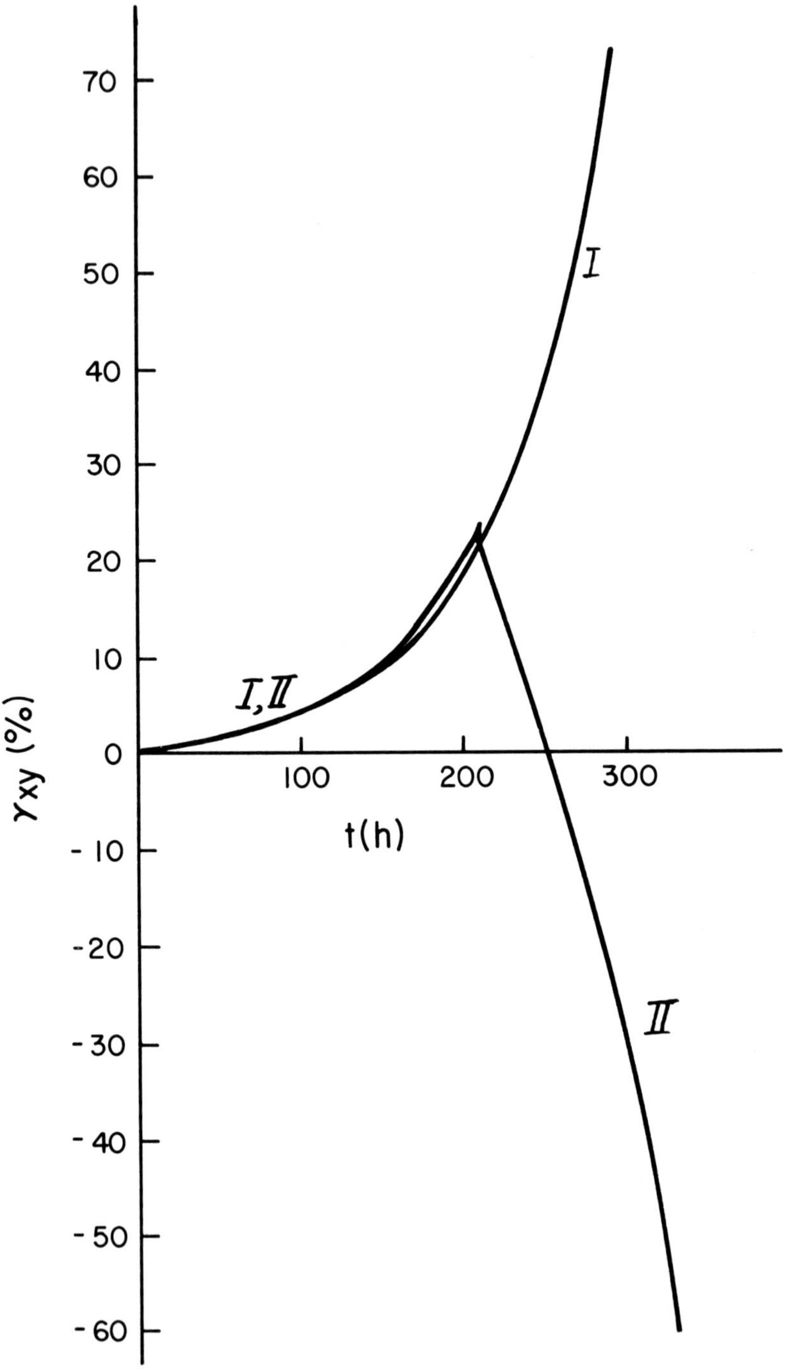

Fig. 2 Reverse torsion test.

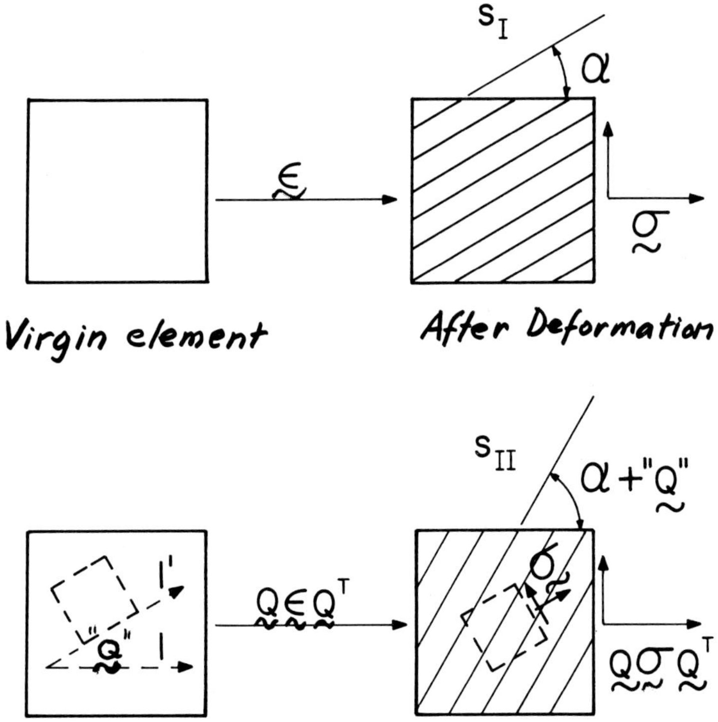

s_I
α
ϵ
σ
Virgin element
After Deformation
s_{II}
α+"Q"
1'
1
"Q"
$Q \epsilon Q^T$
σ
$Q \sigma Q^T$

Fig. 3

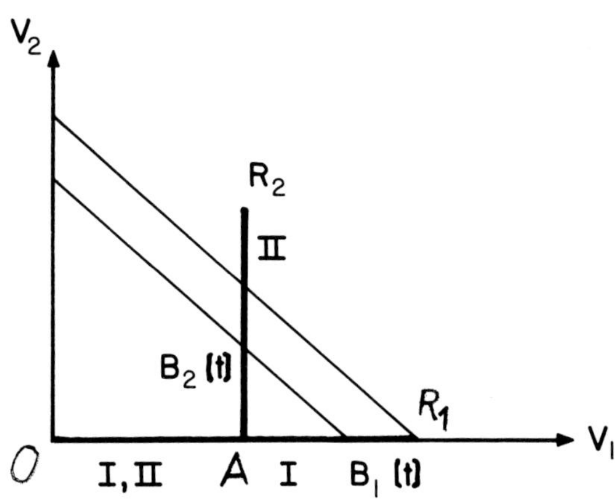

V_2
R_2
II
B_2 (t)
R_1
V_1
O
I, II
A
I
B_1 (t)

Fig. 4

On the Effect of Physical Non-Linearities in the Aeroelastic Stabilitiy Problem of Supersonic Panels

LIVIU LIBRESCU

Department of Solid Mechanics,
School of Engineering,
Tel-Aviv University,
Ramat-Aviv, Israel

Abstract

The problem of the influence of physical non-linearities in determining the dangerous and undangerous character of the critical flutter boundary of supersonic panels is analysed. Practical recommendations concerning the reduction of destabilizing effects exerted by the soft materials are presented.

1. Preliminaries

The new demands for more and still more advanced aircraft and spacecraft vehicles has resulted in the absolute necessity for a more comprehensive approach of their flutter characteristics. In this sense, the non-linear approach of the flutter phenomenon is of interest in two respects. It enables one to determine the conditions in which, due to the character of the considered non-linearities, the critical flutter boundary (determined in the framework of the linear approximation) may be exceeded without an immediate failure of the elastic structure, as well as the conditions in which undamped vibrations may occur at velocities below the flutter critical ones.

According to Bautin's results [1], the above two instances may be reduced to the study of the undangerous and dangerous character of the flutter critical boundary, respectively. The flutter boundary or portions of it are dangerous or undangerous whether $L(\nu_*) > 0$ or $L(\nu_*) < 0$, respectively, where $L(\nu_*)$ denotes the first Liapunov magnitude determined on the flutter boundary. On the other hand, the new and exotic materials used in the construction of aerospace vehicles and working at high temperatures and extreme dynamic pressures, can not be further considered from the viewpoint of their constitutive law as physically linear ones.

Some results concerning the effect played by the physical non-linearities in determining the character of the flutter boundary will be now presented.

2. Governing Equations

The system considered is a flat, thin panel of infinite span and finite length placed in a supersonic gas stream.

The panel (of thickness h) is assumed symmetrically built up of $2\ell + 1$ ($\ell = 1, 2, \ldots$) elastic, isotropic laminae, of constant thickness.

The mid-plane of the panel is referred to a rectangular Cartesian coordinate system (ξ_1, ξ_2), where $\xi_1 = o; a$, define its leading and trailing edge, respectively. The free stream velocity vector is assumed parallel to the $0\xi_1$ axis. It is also assumed that the material of each j-th layer obeys a non-linear physical law; let us adopt the one of Kauderer [2].

The starting point is constituted by the aeroelastic differential equation governing the problem (see [3, 4]), symbolically written as $S(w) - A(w) - I(w) = 0$, where S, A, I stand for the operators: structural (of the physically non-linear panel), aerodynamical (based on the third-order piston aerodynamics) and inertial, respectively; $w \equiv w(\xi_1, t)$ denotes the deflection function expressed as $w(\xi_1, t) = \sum_r w_r(t) f_r(\xi_1)$, where t denotes time, while $f_r(\xi_1)$ are functions fulfilling the boundary conditions on $\xi_1 = o, a$.

Application of Galerkin's method yields a system of coupled non-linear differential equations in time, expressed as:

$$\ddot{\zeta}_r + \varepsilon_T \dot{\zeta}_r + \omega_r^2 \zeta_r + F_r^{(I)}(\zeta_t) + F_r^{(II)}(\zeta_t, M) = 0 \tag{1}$$

$$(\dot{\zeta}_r \equiv d\zeta_r/d\tau, \; (r = 1, 2 \ldots))$$

where $F_r^{(I)}$ and $F_r^{(II)}$ denote respectively the physical and aerodynamical non-linearities, given explicitely in [4]. In addition, for the inserted dimensionless quantities we have:

$$\zeta_r \equiv w_r/h;\ \omega_r^2 \equiv \Omega_r^2/\Omega^2 = r^4;\ \Omega^2 = \pi^4 D_1/(m_o a^4);\tau \equiv \Omega t$$

(the non-dimensional time); M - the Mach number; m_o - the reduced mass of the structure; D_1 - the bending rigidity of the associated linearly-elastic panel; ε_T - the total damping factor.

By considering as large two successive coordinates, ζ_p, ζ_{p+1}, (p = 1, 2,...), only, the so modified governing equations enable one to obtain the first Liapunov magnitude $L(\nu_*)$ under a closed form and afterwards to express the conditions of undangerousness and dangerousness of the flutter boundary. For the case of simply supported panel on $\xi = 0$; a, and of the flow on both faces of the panel with the velocity $U_\infty^+ = U_\infty^- \equiv U$, the former condition reads:

$$C - \nu_*^3 B > 0 \ ,(p = 1,\ 2,\dots) \tag{2}$$

while for the latter one, the inverse inequality is to be considered. In (2)

$$C \equiv K((p+1)^8 - p^8) \text{ and } B \equiv \frac{8}{3}(P\pi^2)(\kappa+1)\ p(p+1)x$$

$$x\ [\frac{p^2(3p^2+2p-1)}{(4p+1)(4p^2-1)} + \frac{(1+p)^2(3p^2+8p+4)}{(2p+1)(4p+3)(2p+3)}] \tag{3}$$

identify the physical and aerodynamical non-linearities, respectively; $\nu_*(\equiv(M\frac{h}{a})_*)$ - the critical flutter speed; κ - the polytropic coefficient;
$$K \equiv \pi^8 h^2 D_2/(a^8 m_o \Omega^2);\ D_2 \equiv \frac{32}{45}\sum_{j=1}^{\ell+1} G^{(j)}(\zeta^3_{(j)} - \zeta^3_{(j-1)})\ x$$

$x\ [2\gamma_2^{(j)}(\mu_{(I)}^{(j)})^2 + \chi_2(j)\ \mu_{(II)}^{(j)})^2]$ - the non-linear bending rigidity of the panel. The sub - or superscript (j) identifies the quantities associated to the jth layer; γ_2 and χ_2 - shearing and stretching coefficients depending on the non-linear physical properties of the material. Their sign, (positive or negative) corresponds to the hard or soft non-linear type materials (as defined in [2]); G - stands for the inplane shear modulus; $\mu_{(I)}$ and $\mu_{(II)}$ - coefficients (see [4]) depending on Poisson's ratio μ_o; $\zeta_{(j-1)}$ the distance from the mid-plane to the interface between (j-1)th and j^{th} panel layers.

3. Conclusions

The inequality (2) and its inverse show the strong influence played by the hard and soft non-linear type materials in determining the character of the flutter boundary (or of portions of it).

Thus, for sufficiently hard type materials, when (2) is fulfilled, the flutter boundary is undangerous.

With the increase of the Mach number, when the aerodynamical non-linearities become prevalent, the flutter boundary turns out to be a dangerous one, in spite of the possible presence of hard materials into the structure. For soft non-linear materials, the flutter boundary becomes a dangerous one. However, in order to maintain the undangerous character of the flutter critical boundary, inclusion of laminae with high non-linear characteristics of the hard type is to be performed. This inclusion may counteract the destabilizing effects of both aerodynamical and of soft physical non-linearities.

References

[1] Bautin, N.N. Behaviour of Dynamic Systems in the Vicinity of the Boundary of the Stability Domain, Gostekizdat, Moskow, 1949.

[2] Kauderer, H., Nichtlineare Mechanik, Springer-Verlag, 1958.

[3] Librescu, L., Aeroelastic Stability of Orthotropic Heterogeneous Thin Panels in the Vicinity of the Flutter Critical Boundary, (Parts I and II), J. Mecanique, vol. 4, 1, (1965). 51-76 and vol. 6, 1, (1967), 133-152.

[4] Librescu, L., Elastostatics and Kinetics of Anisotropic and Heterogeneous Shell-Type Structures, Noordhoff Internat, Publishing, Leyden, The Netherlands 1975.

Plasticity and Ultimate Strength of Concrete

J. LUBLINER

Department of Civil Engineering
University of California
Berkeley, California 94720 U.S.A.

Summary

It is shown that the ultimate strength of concrete under multiaxial compression may be derived from a simple theory of plasticity developed by the author. An example is included.

1. Introduction

The purpose of this note is to examine the ultimate strength of concrete as an application of plasticity theory.

The plasticity theory I have in mind is not the conventional theory, in which the yield criterion plays a central role, but rather the "simple" theory in which plasticity is defined as loading-unloading irreversibility [1]. The main hypothesis is that in the control (e.g., strain-temperature) space there exist "loading" and "unloading" directions such that rate-independent inelastic deformation occurs only upon loading. The short-time compressive behavior of concrete conforms to this hypothesis in a first approximation. While it is true that, upon unloading from an extensively microcracked state, there occurs some reversed inelastic deformation, this appears to be a rate effect.

The tensile behavior of concrete, on the other hand, is classically brittle in being virtually elastic up to fracture. Thus, it may be reasoned that at least two independent failure mechanisms exist in concrete, corresponding respectively to brittle fracture in tension, and slip rupture in shear (resulting from unequal compressive stresses). It is expected that plasticity theory will apply only to processes in which shear dominates, including biaxial compression.

It will be shown that, when the rate equations of plasticity are given, the ultimate-strength condition results from these equations and need not be postulated (as would a yield condition, for example).

2. Theory

When an inelastic material is modelled as a homogeneous continuum, then a local state may be given by the strain tensor $\underset{\sim}{\varepsilon}$, the temperature T, and a number of internal variables $\alpha_1,..,\alpha_n$. In the present model these variables will consist of the plastic strain tensor $\underset{\sim}{\varepsilon}^p$ and a variable κ which measures cumulative plastic deformation and may represent an internal process such as microcracking. It can be shown [2] that under appropriate assumptions there exists a scalar function $F(\underset{\sim}{\varepsilon}, T, \underset{\sim}{\varepsilon}^p, \kappa)$ (the loading function) and a tensor-valued function $\phi(\underset{\sim}{\varepsilon}, T, \underset{\sim}{\varepsilon}^p, \kappa)$ such that

$$\dot{\underset{\sim}{\varepsilon}}^p = \phi\langle\frac{\partial F}{\partial \underset{\sim}{\varepsilon}}:\dot{\underset{\sim}{\varepsilon}} + \frac{\partial F}{\partial T}\dot{T}\rangle \tag{1}$$

$$\kappa = \sqrt{(\dot{\underset{\sim}{\varepsilon}}^p:\dot{\underset{\sim}{\varepsilon}}^p)}. \tag{2}$$

The introduction of a weak stability postulate such as that of Nguyen and Bui [3] leads further to the flow rule

$$\underset{\sim}{\phi} = g\, \underset{\sim}{C}^{-1} \frac{\partial F}{\partial \underset{\sim}{\varepsilon}},$$

where $g(\underset{\sim}{\varepsilon}, T, \underset{\sim}{\varepsilon}^p, \kappa)$ is another scalar function, and $\underset{\sim}{C}$ is the elasticity tensor, assumed positive definite. Thus the plastic strain rate, when it does not vanish, is given (with $\dot{T}=0$) by

$$\dot{\underset{\sim}{\varepsilon}}^p = g\, (\underset{\sim}{C}^{-1} \frac{\partial F}{\partial \underset{\sim}{\varepsilon}})(\frac{\partial F}{\partial \underset{\sim}{\varepsilon}}:\dot{\underset{\sim}{\varepsilon}}). \tag{3}$$

A state $(\underset{\sim}{\varepsilon}, T, \underset{\sim}{\varepsilon}^p, \kappa)$ is an <u>ultimate state</u> if the stress tensor $\underset{\sim}{\sigma}$ is stationary, that is if $\dot{\underset{\sim}{\sigma}} = \underset{\sim}{C}(\dot{\underset{\sim}{\varepsilon}} - \dot{\underset{\sim}{\varepsilon}}^p) = 0$ with $\dot{\underset{\sim}{\varepsilon}} \neq 0$. Since $\underset{\sim}{C}$ is positive definite, this occurs if and only if $\dot{\underset{\sim}{\varepsilon}} = \dot{\underset{\sim}{\varepsilon}}^p$, that is, if

$$g\, C^{-1}_{ijkl} \frac{\partial F}{\partial \varepsilon_{kl}} \frac{\partial F}{\partial \varepsilon_{mn}} \dot{\varepsilon}_{mn} = \dot{\varepsilon}_{ij},$$

and, for $\dot{\underset{\sim}{\varepsilon}} \neq 0$, this requires

$$f(\underset{\sim}{\varepsilon}, T, \underset{\sim}{\varepsilon}^p, \kappa) = 1, \qquad (4)$$

where

$$f \overset{\text{def}}{=} g\, C^{-1}_{ijkl} \frac{\partial F}{\partial \varepsilon_{ij}} \frac{\partial F}{\partial \varepsilon_{kl}} .$$

3. Application

Equation (4) defines a hypersurface in state space. The practical application of ultimate strength, however, demands that it be described in stress space. To that end it is necessary to integrate the rate equations (2), (3) along a given path in stress space, so that the full state at each point of the path is known.

In order to test the applicability of the method, a very simple isotropic model was assumed. First of all, it was assumed that the plastic deformation is distortional only; this is more or less valid for concrete at state up to a point very close to ultimate, but beyond that there is a marked increase in volume resulting from microcrack propagation. The loading function F was assumed (with temperature dependence not shown explicitly) as

$$F(\underset{\sim}{\varepsilon}, \underset{\sim}{\varepsilon}^p, \kappa) = G\, (\underset{\sim}{\varepsilon}' - \underset{\sim}{\varepsilon}^p):(\underset{\sim}{\varepsilon}' - \underset{\sim}{\varepsilon}^p),$$

where $\underset{\sim}{\varepsilon}'$ denotes the strain deviator, and G is the shear modulus. Thus $\frac{\partial F}{\partial \underset{\sim}{\varepsilon}} = \underset{\sim}{\sigma}'$, and $f = \frac{g}{G}\, \underset{\sim}{\sigma}':\underset{\sim}{\sigma}'$; consequently, the ultimate strength criterion may be written formally as

$$\surd(\tfrac{1}{2}\underset{\sim}{\sigma}':\underset{\sim}{\sigma}') = k(\underset{\sim}{\varepsilon}, T, \underset{\sim}{\varepsilon}^p, \kappa),$$

where $k = \surd(G/2g)$. Hence assuming k is equivalent to assuming g.

The rate equations now are

$$\dot{\boldsymbol{\varepsilon}}^p = \frac{\boldsymbol{\sigma}' \langle \boldsymbol{\sigma}' : \dot{\boldsymbol{\varepsilon}} \rangle}{k\sqrt{(2\,\boldsymbol{\sigma}' : \boldsymbol{\sigma}')}} , \qquad (5)$$

$$\dot{\kappa} = \frac{\langle \boldsymbol{\sigma}' : \dot{\boldsymbol{\varepsilon}} \rangle}{k} . \qquad (6)$$

The following form was assumed for k:

$$k = k_o - Ga\kappa - b(3K \operatorname{tr} \boldsymbol{\varepsilon}) + c(3K \operatorname{tr} \boldsymbol{\varepsilon})^2/k_o .$$

based on the model of Bresler and Pister [4]. Here G and K are the shear and bulk moduli, and a, b, c and k_o are constants.

We now consider an isothermal process starting from (0, T, 0, 0) at time t = 0, such that $\boldsymbol{\sigma}(t) = -\boldsymbol{\mu}\,\sigma(t)$, where $\sigma(t)$ is the numerically largest principal stress at time t, and $\boldsymbol{\mu}$ is a constant tensor with $M_1 \overset{\text{def}}{=} \operatorname{tr} \boldsymbol{\mu} \geq 1$. Let $\boldsymbol{\mu}'$ denote the deviator of $\boldsymbol{\mu}$ and define

$$M_2 = \sqrt{(\tfrac{1}{2}\boldsymbol{\mu}' : \boldsymbol{\mu}')}, \quad m = M_1/M_2 .$$

It follows from equations (5) and (6) that $\boldsymbol{\varepsilon}^p(t) = -\boldsymbol{\mu}'\kappa(t)/2M_2$, hence

$$\boldsymbol{\varepsilon}' = -\boldsymbol{\mu}'\gamma(t)/2M_2$$

where

$$\gamma = \sqrt{(2\boldsymbol{\varepsilon}' : \boldsymbol{\varepsilon}')} = \kappa + M_2\sigma/G .$$

Thus equation (6) becomes $\kappa = M_2\sigma\,\dot{\gamma}/k$. Upon defining $x = G\gamma/k_o$, $y = M_2\sigma/k_o$, we obtain the differential equation

$$\frac{dy}{dx} = 1 - \frac{y}{1 - ax + (a + bm)y + cm^2y^2} .$$

With assumed values of a, bm and cm^2, this equation can be solved numerically so that an effective stress-strain curve is obtained. The point where dy/dx = 0 represents an ultimate state. Let the value of y at this point be denoted $y_m(a, bm, cm^2)$. In a biaxial compression test with $\sigma_1 = -\sigma$, $\sigma_2 = -r\sigma$ $(0 \leq r \leq 1)$, $\sigma_3 = 0$, we have $m = \sqrt{3}(1 + r)/\sqrt{(1 - r + r^2)}$. Thus for assumed values of a, b and c we can construct the biaxial failure curve, given parametrically by

$$\frac{\sigma_1}{k_o} = -\sqrt{\left(\frac{3}{1 - r + r^2}\right)}\, y_m(a, bm, cm^2)$$

$$\frac{\sigma_2}{k_o} = -r\sqrt{\left(\frac{3}{1 - r + r^2}\right)}\, y_m(a, bm, cm^2).$$

The uniaxial ultimate stress corresponds to r = 0.

It is now possible to use optimization techniques to find values of a, b and c so as to produce a least-squares fit with empirical results for various concretes. One example is shown in Fig. 1, where the experimental data are those of Kupfer et al. [5].

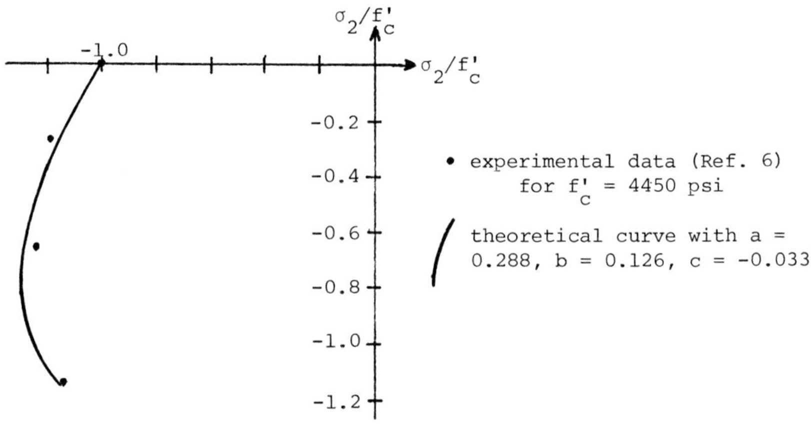

Figure 1

References

1 Lubliner, J.: A simple theory of plasticity. Int. J. Solids & Structures 10 (1974) 313.

2 Lubliner, J.: On loading, yield and quasi-yield hypersurfaces in plasticity theory. Int. J. Solids & Structures 11 (1975) 1011.

3 Nguyen, Q.S.; Bui, H.D.: Sur les matériaux élastoplastiques à écrouissage positif ou négatif. J. de Mec. 13 (1974) 321.

4 Bresler, B.; Pister, K.S.: Failure of plain concrete under combined stresses. Trans. ASCE 122 (1957) 1049.

5 Kupfer, H.; Hilsdorf, H.; Rusch, H.: Behavior of concrete under biaxial stress. ACI Journal 66 (1969) 656.

An Internal Variable Approach to the Formulation of Finite Element Problems in Plasticity

J.B. MARTIN

Department of Civil Engineering
University of Cape Town
Rondebosch 7700, South Africa.

Summary

The paper is a brief discussion on the potential advantages of employing an internal variable framework for the conceptual formulation of finite element problems in plasticity. The discussion is carried out primarily in terms of the small displacement rate problem for structures with piecewise linear yield functions, where a comparison can be made between the internal variable formulation and the programming formulation proposed by G. Maier and his co-workers. Implications of this comparison are briefly discussed.

1. Introduction

Internal variable theories, with a sound thermodynamic basis, have been used increasingly in recent years to provide a framework for the formulation of constitutive equations in plasticity. Most frequently these applications have been on the scale of infinitesimal continuum elements, with the internal variables conceived as describing micro-structural rearrangements. The constitutive equations so derived are then combined with the appropriate continuum equations of kinematics and equilibrium or dynamics to complete the formulation of a structural problem.

However, the internal variable framework can be applied directly to finite elements, with the internal variables now associated with rearrangements within the elements. This procedure is quite simple and straightforward, but it offers the potential of new insights into structural behaviour and the conceptual formulation of finite element problems in plasticity, and as such is deserving of further exploration. In view of the limitations on space, we shall choose to illustrate this potential by considering in detail one particular problem.

This problem will be the isothermal, small displacement rate or incremental problem for elastic-plastic bodies subjected to conservative loading. The problem will be formulated, in discrete terms, within an internal variable framework which is essentially that of Kestin and Rice [1], in the form given by Martin [2]. The governing equations will be derived and the most directly accessible kinematic minimum principle will be set up.

We shall see that in this process, with the particular assumptions made, we duplicate the formulation given by Maier [3]. Maier obtained his result from a conventional mechanical formulation of the governing equations and the recognition that these equations were the Kuhn-Tucker conditions for a quadratic programming problem. To assist in the comparison of the formulations, we shall use the same symbols as are adopted in the recent survey on mathematical programming methods in engineering plasticity [4].

Several interesting points arise from the comparison of the two formulations. In one direction, relationships between quantities used in Maier's formulation are suggested. In the other direction, Maier's intuitive formulation of hardening rules suggests that simple mechanical models can be constructed for conventional hardening rules. Further, the overall structure of the internal variable approach leads to some conjectures regarding the manner in which finite elements problems in plasticity may be formulated.

2. The Elastic Energy of an Element

Consider an *element* of a structure; this element may be an infinitesimal element or a conventionally defined finite element. Let the natural element strains be $\underline{q}$, and the conjugate generalised stresses be $\underline{Q}$. Linear elastic isothermal behaviour can be characterised by a stiffness matrix $\underline{k}$ which is symmetric and positive definite. The free energy, or elastic strain energy, is

$$f = \tfrac{1}{2}\,\underline{q}^T\,\underline{k}\,\underline{q} \quad \text{or} \quad f = \tfrac{1}{2}\,\underline{q}^T\,\underline{Q}\,, \tag{1}$$

which provides the elastic constitutive relations

$$\underline{Q} = \underline{k}\,\underline{q} \quad \text{or} \quad \underline{q} = \underline{k}^{-1}\,\underline{Q} \,. \tag{2}$$

Plastic behaviour is described by the addition of a vector of internal variables $\underline{\lambda}$. The free energy is now a function of $\underline{q}$ and $\underline{\lambda}$, and we shall assume that it is homogeneous and quadratic in its arguments. Thus

$$f = \tfrac{1}{2}\left\{\begin{matrix} q \\ \lambda \end{matrix}\right\}^T \underline{k}^* \left\{\begin{matrix} q \\ \lambda \end{matrix}\right\} = \tfrac{1}{2}\left\{\begin{matrix} q \\ \lambda \end{matrix}\right\} \begin{bmatrix} k & -A \\ -A^T & C \end{bmatrix} \left\{\begin{matrix} q \\ \lambda \end{matrix}\right\}. \tag{3}$$

In this equation $\underline{k}^*$ is assumed to be symmetric, and f reduces to equn. (1) when $\underline{\lambda} = 0$. The submatrices $\underline{A}$ and $\underline{C}$ are yet to be defined. Let the internal forces associated with the $\underline{\lambda}$ be $-\underline{X}$, so that the free energy may be written as

$$f = \tfrac{1}{2}\,\underline{q}^T\,\underline{Q} + \tfrac{1}{2}\,\underline{\lambda}^T\,(-\underline{X}) \,. \tag{4}$$

It follows that

$$\underline{Q} = \underline{k}\,\underline{q} - \underline{A}\,\underline{\lambda} \,, \quad \underline{X} = \underline{A}^T\,\underline{q} - \underline{C}\,\underline{\lambda} \quad . \tag{5}$$

From the first of these equations, we may write

$$\underline{q} = \underline{k}^{-1}\,\underline{Q} + \underline{k}^{-1}\,\underline{A}\,\underline{\lambda} = \underline{e} + \underline{p} \quad , \tag{6}$$

where $\underline{e}$, $\underline{p}$ are by definition the *elastic* and *plastic* components of strain respectively. Hence

$$\underline{e} = \underline{k}^{-1}\,\underline{Q}, \quad \underline{p} = \underline{N}\,\underline{\lambda} \quad \text{where } \underline{N} = \underline{k}^{-1}\,\underline{A}. \tag{7}$$

It follows from equns. (6) and (7) that

$$\underline{e} = \underline{q} - \underline{p} = \underline{q} - \underline{N}\,\underline{\lambda}, \quad \text{and} \quad \underline{A} = \underline{k}\,\underline{N} \,. \tag{8}$$

The matrix $\underline{k}^*$ may be rewritten as

$$\underline{k}^* = \begin{bmatrix} k & -A \\ -A^T & C \end{bmatrix} = \begin{bmatrix} k & -kN \\ -N^T k & N^T kN+H \end{bmatrix}, \tag{9}$$

where $\underline{H}$ is symmetric but still undefined. Substitution of equn. (9) into equn. (3) will show that we may write

$$\begin{aligned} f &= \tfrac{1}{2}(\underline{q} - \underline{N}\,\underline{\lambda})^T\,\underline{k}(\underline{q} - \underline{N}\,\underline{\lambda}) + \tfrac{1}{2}\,\underline{\lambda}^T\,\underline{H}\,\underline{\lambda} \\ &= \underline{e}^T\,\underline{k}\,\underline{e} + \tfrac{1}{2}\,\underline{\lambda}^T\,\underline{H}\,\underline{\lambda} \quad . \end{aligned} \tag{10}$$

This equation may be seen as the direct generalisation of equn. (1). From the second of equns. (5),

$$\begin{aligned} \underline{X} &= \underline{N}^T \underline{k}\, \underline{q} - (\underline{N}^T \underline{k}\, \underline{N} + \underline{H})\, \underline{\lambda} \\ &= \underline{N}^T \underline{k}(\underline{q} - \underline{N}\, \underline{\lambda}) - \underline{H}\, \underline{\lambda} \\ &= \underline{N}^T \underline{Q} - \underline{H}\, \underline{\lambda}. \end{aligned} \tag{11}$$

It can thus be seen that the free energy, or strain energy, can be defined in terms of the three element matrices $\underline{k}$, $\underline{N}$, $\underline{H}$. The conventional assumptions that the elastic behaviour is linear, that $\underline{k}$ is symmetric and positive definite and that $\underline{k}^*$ is symmetric have been made, but no further restrictions have been imposed.

3. The Dissipation Function

In addition to the fundamental equation (10), from which all equations describing the elastic behaviour follow, a kinetic equation relating the rates of change of the internal variable rates $\dot{\underline{\lambda}}$ and the internal forces $\underline{X}$ must be introduced. The rate of internal dissipation is

$$D = \dot{\underline{\lambda}}^T \underline{X} . \tag{12}$$

It is directly related to the entropy production rate, and is required by the second law to be non-negative.

For time-independent plasticity it is essential that $\underline{X}$ should be homogeneous and of degree zero in the components in $\dot{\underline{\lambda}}$ (otherwise the kinetic relation is a differential equation through which time is introduced into the constitutive equations). The dissipation function is then homogeneous and of degree one in $\dot{\underline{\lambda}}$. It will also be assumed that the dissipation function is convex; some of the implications of this assumption have been discussed by Martin [5].

For our present purposes, we shall make the further simplifying assumption that there exist independent relations between each component of $\dot{\underline{\lambda}}$ and the associated component of $\underline{X}$. The dissipation function for the i-th component of $\underline{\lambda}_i$, $\dot{\lambda}_i$ will be assumed to be as shown in Fig 1(a), and the relation between X_i and $\dot{\lambda}_i$ as shown in Fig 1(b). In mechanical terms, these diagrams describe a *rigid-plastic* relation. It can be expressed mathematically in a variety of ways. Conventionally, for example, it is written as

$$X_i = R_i \text{ when } \dot{\lambda}_i > 0, \quad X_i \leq R_i \text{ when } \dot{\lambda}_i = 0, \tag{13}$$

with $\dot{\lambda}_i < 0$ not admitted, or as

$$\dot{\lambda}_i \geq 0 \text{ when } X_i = R_i, \; \dot{\lambda}_i = 0 \text{ when } X_i < R_i, \tag{14}$$

with $X_i > R_i$ not admitted. Alternatively, it can be expressed as a complementarity relation (Cohn and Maier [4]), emphasising the sign constraints on $\dot{\lambda}_i$ and $X_i - R_i$;

$$(X_i - R_i)\dot{\lambda}_i = 0, \; X_i - R_i \leq 0, \quad \dot{\lambda}_i \geq 0 \; . \tag{15}$$

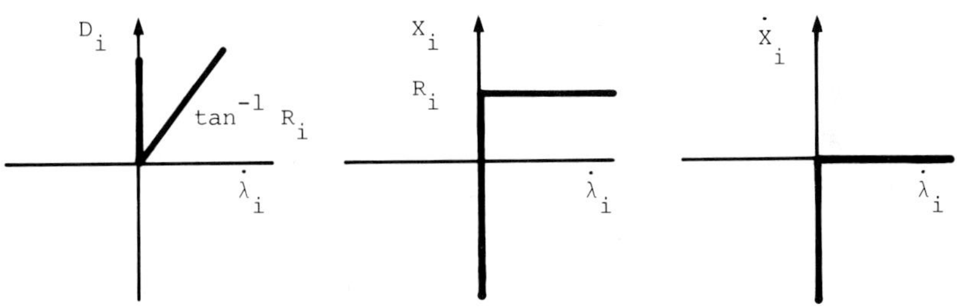

Fig. 1 Relations between X_i and $\dot{\lambda}_i$

It is convenient to form an element vector R, with components R_i, and an element vector $\underline{\phi}$ defined by

$$\underline{\phi} = \underline{X} - \underline{R} = \underline{N}^T \underline{Q} - \underline{H}\,\underline{\lambda} - \underline{R} \; , \tag{16}$$

after using equn. (11). The vector ϕ is recognised as a vector of piecewise linear yield functions. Equns. (14) can now be written for the element as a whole;

$$\underline{\dot{\lambda}}^T \underline{\phi} = 0, \quad \underline{\phi} \leq 0, \quad \underline{\dot{\lambda}} \geq 0 \quad . \tag{17}$$

It can also be argued mechanically that, provided $\phi_i = X_i - R_i = 0$, a relation exists between $\dot{\phi}_i = \dot{X}_i$ and $\dot{\lambda}_i$ of the form shown in Fig 1(c). This relation can also be expressed as a complementarity condition;

$$\dot{\phi}_i \, \dot{\lambda}_i , \quad \dot{\phi}_i \le 0, \quad \dot{\underline{\lambda}} \ge 0 \; . \tag{18}$$

4. Global Formulation

We now order the elements of the structure, and form global strain, generalised stress, internal variable and internal force vectors by writing the element vectors in order. Without confusion we can again denote these vectors by $\underline{q}$, $\underline{Q}$, $\underline{\lambda}$, $\underline{X}$. Similarly we form block diagonal matrices, $\underline{k}$, $\underline{N}$, $\underline{H}$ by assembling the associated element matrices in order. In this way it is clearly seen that the essential relations defined for the element hold for the assembly of elements.

The global vector of unconstrained node displacements is denoted by $\underline{u}$. Consistent with the assumption of small displacements, we introduce a fixed deformation matrix $\underline{C}$ which specifies the manner in which the elements are connected;

$$\underline{q} = \underline{C} \, \underline{u} \; . \tag{19}$$

A nodal load vector $\underline{F}$ is also introduced; the components of $\underline{F}$ are defined so that the potential of the loads is $-\underline{u}^T \underline{F}$.

The free energy (or the potential energy) of the assemblage and the conservative loads applied to it is

$$\begin{aligned} U_p &= \Sigma f - \underline{u}^T \underline{F} \\ &= \tfrac{1}{2}(\underline{C}\,\underline{u} - \underline{N}\,\underline{\lambda})^T \underline{k}(\underline{C}\,\underline{u} - \underline{N}\,\underline{\lambda}) + \tfrac{1}{2}\,\underline{\lambda}^T \underline{H}\,\underline{\lambda} - \underline{u}^T \underline{F} \; . \end{aligned} \tag{20}$$

Since the displacements $\underline{u}$ are unconstrained, a necessary condition for equilibrium is that

$$(\delta U_p)_{\lambda \text{ fixed}} = 0 \quad . \tag{21}$$

This gives

$$\delta \underline{u}^T \underline{C}^T \underline{k}(\underline{C}\,\underline{u} - \underline{N}\,\underline{\lambda}) - \delta\underline{u}^T F = 0 \quad , \tag{22 a}$$

or

$$\underline{C}^T \underline{k} \, (\underline{C}\,\underline{u} - \underline{N}\,\underline{\lambda}) = \underline{F} \quad , \tag{22 b}$$

which is the conventional equilibrium equation.

A global yield function vector $\underline{\phi}$ is now written as

$$\underline{\phi} = \underline{N}^T \underline{Q} - \underline{H}\,\underline{\lambda} - \underline{R} = \underline{N}^T \underline{k}\,\underline{C}\,\underline{u} - (\underline{N}^T \underline{k}\,\underline{N} + \underline{H})\underline{\lambda} - \underline{R}\ , \tag{23}$$

and the constitutive equations are otherwise unchanged;

$$\underline{\dot{\lambda}}^T \underline{\phi} = 0, \quad \underline{\phi} \leq 0, \quad \underline{\dot{\lambda}} \geq 0\ . \tag{24}$$

Equns. (22 b) and (24) characterise the elastic-plastic problem. However, because equn. (24) is of degree zero in $\underline{\dot{\lambda}}$, these equations in $\underline{u}$, $\underline{\lambda}$, $\underline{\dot{\lambda}}$ cannot be solved in their present form. We consider the rate problem in $\underline{\dot{u}}$, $\underline{\dot{\lambda}}$ for a load rate $\underline{\dot{F}}$. The equilibrium equation (22 b) may be differentiated to give

$$\underline{C}^T \underline{k}\,(\underline{C}\,\underline{\dot{u}} - \underline{N}\,\underline{\dot{\lambda}}) = \underline{\dot{F}}\ . \tag{25}$$

The components of the vector $\underline{\lambda}$, λ_i, are now assigned to either of two vectors $\underline{\lambda}_p$, $\underline{\lambda}_e$ according to whether $\phi_i = 0$ or $\phi_i < 0$, respectively. We similarly assign the components of $\underline{\phi}$ to vectors $\underline{\phi}_p$, $\underline{\phi}_e$. Then, from equn. (17), we have

$$\underline{\phi}_p = 0, \quad \underline{\dot{\lambda}}_p^T \underline{\dot{\phi}}_p = 0, \quad \dot{\phi}_p < 0, \quad \dot{\lambda}_p \geq 0 \tag{26 a}$$

and

$$\underline{\phi}_e < 0, \quad \underline{\dot{\lambda}}_e = 0. \tag{26 b}$$

Equns. (25) and (26) characterise the *rate problem.*

We have arrived here at a formulation of the kinematic rate problem for piecewise linear yield surfaces which is quite conventional. It is in exploring some of the implications of the method of formulation that we can appreciate some of the insights which the internal variable formulation provides.

5. Kinematic Minimum Principles

Maier [3] recognised that equns. (25) and (26) were in fact the Kuhn-Tucker conditions for a quadratic programming problem. Following a mechanical argument, as suggested by Carter and Martin [6], we can reproduce this result very simply, and in such a way that some physical insight can be gained.

Considering equns. (22 a) and (22 b), it is evident that equn. (25) can be written as

$$\delta \dot{\underline{u}}^T \underline{C}^T \underline{k} (\underline{C} \dot{\underline{u}} - \underline{N} \dot{\underline{\lambda}}) - \delta \dot{\underline{u}}^T \dot{\underline{F}} = 0 \tag{27}$$

Further, on considering Fig 1(c), it is also evident that we can write

$$\delta \dot{\underline{\lambda}}_p \dot{\underline{\phi}}_p \leq 0 . \tag{28 a}$$

Denoting by $\underline{N}_p$, $\underline{H}_p$ those parts of the matrices $\underline{N}$, $\underline{H}$ which involve the active yield planes for which $\phi_i = 0$, from equn. (23) we find

$$\dot{\underline{\phi}}_p = \underline{N}_p^T \underline{k} \underline{C} \dot{\underline{u}} - (\underline{N}_p^T \underline{k} \underline{N}_p + \underline{H}_p) \dot{\underline{\lambda}}_p , \tag{28 b}$$

and hence equn. (28a) becomes

$$\delta \dot{\underline{\lambda}}_p \underline{N}_p^T \underline{k} \underline{C} \dot{\underline{u}} - \delta \dot{\underline{\lambda}}_p (\underline{N}_p^T \underline{k} \underline{N}_p + \underline{H}_p) \dot{\underline{\lambda}}_p \leq 0 . \tag{28 c}$$

Subtracting equn. (28c) from (27), we find

$$\delta\{\tfrac{1}{2}(\underline{C} \dot{\underline{u}} - \underline{N}_p \dot{\underline{\lambda}}_p)^T \underline{k} (\underline{C} \dot{\underline{u}} - \underline{N}_p \dot{\underline{\lambda}}_p) + \tfrac{1}{2} \dot{\underline{\lambda}}_p^T \underline{H} \dot{\underline{\lambda}}_p - \dot{\underline{u}}^T \underline{F}\} \geq 0$$

$$\text{subject to } \dot{\underline{\lambda}}_p \geq 0, \quad \dot{\underline{\lambda}}_e = 0. \tag{29}$$

This leads directly to Maier's quadratic programming problem in $\dot{\underline{u}}$, $\dot{\underline{\lambda}}$ for the solution of the kinematic rate problem.

Elimination of $\dot{\underline{\lambda}}$ from the governing equations (25) and (26) leads to the conventional tangent stiffness method. The classical minimum principle (Greenberg [7]) may be obtained if we note that

$$\begin{aligned} \underline{\lambda}_p^T \dot{\underline{\phi}}_p &= \dot{\underline{\lambda}}_p \underline{N}_p^T \underline{Q} - \dot{\underline{\lambda}}_p^T \underline{H} \dot{\underline{\lambda}}_p = (\underline{N}_p \dot{\underline{\lambda}}_p)^T \dot{\underline{Q}} - \dot{\underline{\lambda}}_p^T \underline{H} \dot{\underline{\lambda}}_p \\ &= \dot{\underline{p}}^T \dot{\underline{Q}} - \dot{\underline{\lambda}}_p \underline{H} \dot{\underline{\lambda}}_p = 0. \end{aligned} \tag{30}$$

$$\text{Since } (\underline{C} \dot{\underline{u}} - \underline{N}_p \dot{\underline{\lambda}}_p)^T k (\underline{C} \dot{\underline{u}} - \underline{N}_p \dot{\underline{\lambda}}_p) = \dot{\underline{e}}^T \dot{\underline{Q}} , \tag{31}$$

equn. (29) can be written as

$$\delta \{\tfrac{1}{2} \dot{\underline{q}}^T \dot{\underline{Q}} - \dot{\underline{u}}^T \dot{\underline{F}}\} \geq 0 \quad (32)$$

subject to $\dot{\underline{q}} = \underline{C}\,\dot{\underline{u}}$, it being understood that $\dot{\underline{Q}}$ is the stress rate associated with $\dot{\underline{q}}$ through the constitutive relations. Equn. (32) leads directly to Greenberg's minimum principle; this simple extension follows the argument given in Cohn and Maier [4], Chapter 13.

In summary, the basic minimum principle is very simply formulated as an extension of the thermodynamic equilibrium condition (27). Alternative forms may similarly be directly generated.

6. Constitutive Relations

If we return to the element level, and consider the case of uniaxial tension and compression, the internal variable framework under discussion implies that a mechanical model made up of linear springs and rigid-plastic sliding elements can be constructed to illustrate any plastic constitutive equation which can be generated by the theory. This is of course well known for elastic, perfectly plastic relations and linear kinematic hardening models, which are shown in Figs 2(a) and 2(b). In these models λ^+ is associated with a rigid-plastic element which can only extend, and λ^- with an element which can only compress. A model for isotropic hardening may not be quite so familiar, although we guess its existence in that a symmetric $\underline{H}$ matrix has been given by Maier [8] for this case. The model is shown in Fig 2(c) without detailed derivation. Note the interesting feature, however, that the stiffness of one spring is negative if $2E_T < E$, as is most likely.

These mechanical models may be conceived of as applying to pairs of yield planes in a more complex element; it is a quite straightforward process to generalise from the $\underline{H}$ matrix for uniaxial tension to the multiaxial case. Thus, provided $\underline{H}$ is symmetric, we can anticipate that an internal variable model can be formulated for any piecewise linear yield surface and hardening rate constructed in mechanical terms, and that we can further construct a simple mechanical model of springs and rigid-plastic elements. A slight difference may occur between the mechanical and the internal variable formulations if plastic multipliers, which are proportional but not necessarily identical to the internal variables, are used as basic variables.

7. A Simple General Formulation

The ideas set out above suggest that a simple general formulation of all small displacement discrete rate or incremental problems in time independent plasticity is possible. Consider a structure subjected to conservative nodal loading $\underline{F}(t)$. Let the unconstrained nodal displacement be $\underline{u}$ and the internal variables be $\underline{\lambda}$ as before. For convenience, however, let f be the free energy of the structure (i.e. the sum of the energies of the elements) and let D be the dissipation function for the structure.

We assume that $f = f(\underline{u}, \underline{\lambda})$ and that it is a homogeneous, quadratic function of its arguments. This assumption, as we have pointed out, is usual in engineering plasticity, in the small displacement case. We shall futher assume that $D = D(\dot{\underline{u}}, \dot{\underline{\lambda}})$; the dependence on $\dot{\underline{u}}$ is added to permit rigid-plastic elements to be added to the structure. It is conjectured that any time independent plastic behaviour adopted in calculations in engineering plasticity can be obtained from a dissipation function which is homogeneous and of degree one in its arguments. Adopting this, we note that

$$D(\dot{\underline{u}}, \dot{\underline{\lambda}}) = \frac{\partial D}{\partial \dot{\underline{u}}} \dot{\underline{u}} + \frac{\partial D}{\partial \underline{\lambda}} \dot{\underline{\lambda}} . \tag{33}$$

If we now write a work rate balance for the isothermal structure (i.e. the first law of thermodynamics), we have

$$\frac{df}{dt} + D = \dot{\underline{u}}^T \underline{F} . \tag{34}$$

This we may write as

$$\dot{\underline{u}}^T \left(\frac{\partial f}{\partial \underline{u}} + \frac{\partial D}{\partial \dot{\underline{u}}} - \underline{F} \right) + \dot{\underline{\lambda}}^T \left(\frac{\partial D}{\partial \dot{\underline{\lambda}}} + \frac{\partial f}{\partial \underline{\lambda}} \right) = 0. \tag{35}$$

In the case where $D = D(\dot{\underline{\lambda}})$, we see in the two groups of terms in parentheses the basic equations of the system. More generally, the

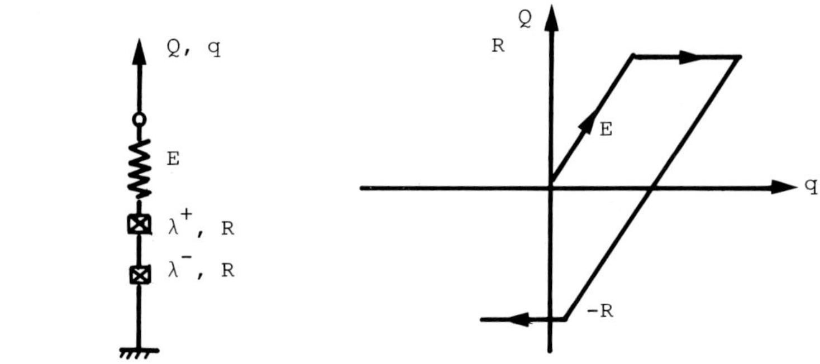

(a) Elastic, perfectly plastic model

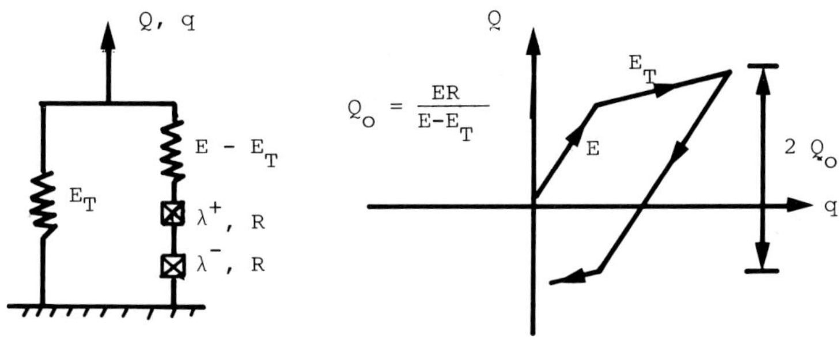

(b) Linear kinematic hardening model

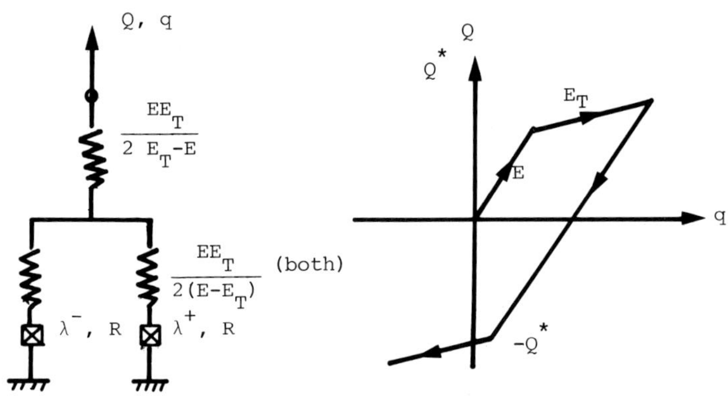

(c) Linear isotropic hardening model

Fig. 2 Mechanical models of plastic behaviour

equations of the system are

$$\underline{C}^T \underline{Q} + \frac{\partial D}{\partial \dot{u}} = F, \quad \underline{X} = \frac{\partial D}{\partial \dot{\lambda}} , \tag{36}$$

using the relations $q = \underline{c}\ \underline{u}$, $\underline{Q} = \partial f/\partial \underline{q}$, $\underline{X} = -\partial f/\partial \underline{\lambda}$. The rate form of the equations can be obtained as before. Following the arguments suggested by Green and Rivlin [8], a rigorous derivation of these equations may be possible.

Conceptually these equations are very simply obtained; they permit time independent plasticity to be easily related to time dependent problems, corresponding to the case where D is not of degree one in the components of $\dot{\underline{u}}$, $\dot{\underline{\lambda}}$. Depending on the choice of the D function we can generate plastic behaviour corresponding to piecewise linear models, or more complex yield surfaces.

References

1 J. Kestin and J.R. Rice, "Paradoxes in the Application of Thermodynamics to Strained Solids", A Critical Review of Thermodynamics, edited by E.B. Stuart, B. Gal-or and A.J. Brainard, Mono Book Corp., 275-298, 1970.

2 J.B. Martin, Plasticity, M.I.T. Press, 1975.

3 G. Maier, "Some Theorems for Plastic Strain Rates and Plastic Strains", J. de Mécanique, 8 , 5-19, 1969.

4 M.Z. Cohn and G. Maier (editors), Engineering Plasticity by Mathematical Programming, Pergamon Press, 1979.

5 J.B. Martin, "A note on the implications of thermodynamic stability in the internal variable theory of inelastic solids", Int. J. Solids and Structures, 11 , 247-253, 1975.

6 P. Carter and J.B. Martin, "A note on the derivation of the kinematic rate theorem of plasticity from the free energy minimum principle", J. de Mécanique, 16 , 791-802, 1977.

7 H.J. Greenberg, "Complementary minimum principles for an elastic plastic material", Quart. Appl. Math., 7 , 85-95, 1949.

8 G. Maier, "Piecewise linearization of yield criteria in structural plasticity", SM Archives, 1 , 239-281, 1976.

9 A.E. Green and R.S. Rivlin, "On Cauchy's equation of motion", ZAMP, 15, 290-292, 1964.

Sur la Consolidation Post-Fluage des Materiaux Metalliques Soumis à un Etirement Progressif

R. MAZET

Professeur Honoraire des Universités
38, Rue Madeleine Cresnon

92320 SCEAUX (FRANCE)

Summary.

The concept of "plastic rate" of an heteroplastic material is applied to an engineering metal submitted to a progressive tension in order to point out an increase of elasticity following the total plastification. This "consolidation" small in quantity plays an important role.

I. Une modélisation simple du comportement d'un matériau métallique soumis à un étirement progressif consiste à définir le taux de plastification θ comme le rapport, dans un élément de volume, du nombre des micro-éléments plastifiés au nombre total des micro-éléments. Dans une éprouvette cylindrique idéale soumise à une traction simple, θ dépend au plus de l'allongement $\varepsilon = \mathrm{Ln}(l/l_o)$ et de la vitesse d'allongement $\dot{\varepsilon}$. Un modèle particulièrement simple et souvent employé est celui du "matériau élastique parfaitement plastique" dans lequel θ passe instantanément de 0 à 1 quand ε franchit la limite élastique ε_G (Fig.1). Un tel matériau peut être appelé homoplastique.

2. Malheureusement, lorsqu'il s'agit d'un matériau métallique du commerce (métal ou alliage) non traité en laboratoire en vue de faire disparaître les dislocations, la réalité est plus complexe :

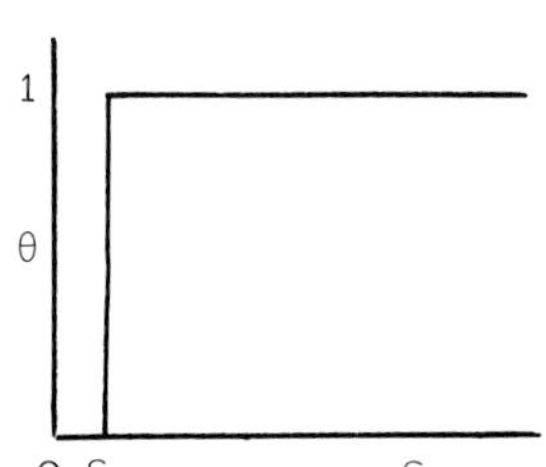

FIG. I

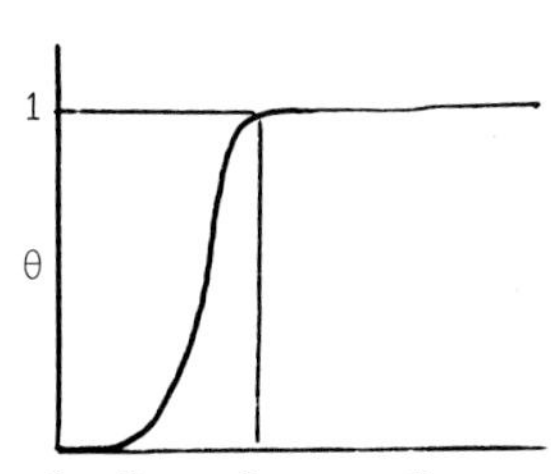

FIG. II

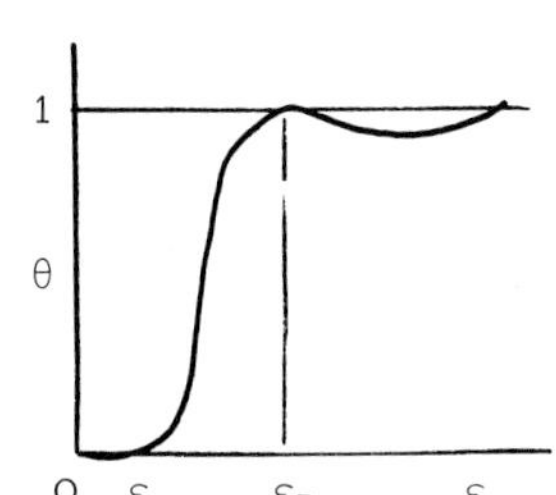

FIG. III

chaque élément (polycristal) comprend un grand nombre de micro-éléments enchevêtrés, différemment orientés et ayant des limites élastiques différentes dans la direction de la sollicitation d'ensemble x'x. Il en résulte que θ passe progressivement de 0 à 1 quand ε parcourt un intervalle, généralement petit, compris entre ε_G(limite du domaine réversible) et ε_L(seuil du fluage généralisé) (Fig.2). Le matériau est dit alors <u>hétéroplastique</u>.

3. Pour l'éprouvette cylindrique idéale, il est facile d'établir la relation qui relie la contrainte macroscopique σ à ε et à θ en admettant les hypothèses suivantes (i désigne un micro-élément, ε_i^e et ε_i^p respectivement sa déformation élastique et sa déformation plastique dans la direction x'x) :

a) $\sigma_i = E\,\varepsilon_i^e = EK(\dot{\varepsilon}^p)^{1/n}$ conformément aux lois de HOOKE et de NORTON (E:module d'YOUNG, K et n : constantes spécifiques du matériau).

b) Si $\dot{\varepsilon} = \text{Const}^{te}$: $\dot{\varepsilon}_i = \dot{\varepsilon}$

Dans le cas particulier b) auquel nous nous limiterons ici et si n est assez grand - ce qui est le cas des métaux et alliages usuels aux températures modérées (moins de 300°C) - la relation prend la forme simple [1] .

$$\frac{\sigma}{E} = k = (1-\theta)^{\varepsilon} + \int_{\mathrm{Sup}(0,\,\varepsilon-\lambda)}^{\varepsilon} \theta(u)\,du \qquad (1)$$

avec $\lambda = K(\dot{\varepsilon})^{1/n}$

λ est la "mémoire" du matériau étiré à la vitesse <u>constante</u> $\dot{\varepsilon}$.

La formule (1) redonne, pour $\varepsilon \lessdot \varepsilon_G$, la loi de HOOKE $k=\varepsilon$ pour le domaine élastique et, pour ε infini (à interpréter physiquement), la loi de BAILEY pour le fluage établi :

$k = \lim_{\varepsilon \to \infty} \left[(1-\theta)\varepsilon\right] + \lambda \simeq K(\dot{\varepsilon})^{1/n}$ si σ est suffisamment grand.

4. La relation (1) permet de calculer la fonction $\theta(\varepsilon \mid \dot{\varepsilon})$ en y introduisant au premier membre la fonction $k(\varepsilon \mid \dot{\varepsilon})$ déduite, pour un matériau donné, d'un essai de traction simple à la vitesse constante $\dot{\varepsilon}$. Bien qu'elle ait été établie pour une éprouvette idéale, les renseignements qu'elle fournit peuvent être considérés comme valables, au moins à titre indicatif, pour l'éprouvette étudiée. Or l'expérience montre <u>toujours</u>, s'agissant de matériaux métalliques usuels, que la courbe θ affecte l'allure

de la Fig. 3 : au delà de ε_L, θ ne conserve pas la valeur 1, mais diminue d'une faible quantité (moins de 1%), puis, si le matériau est suffisamment ductile, croit de nouveau vers 1 jusqu'à ce que la rupture intervienne. Ce comportement peut s'interpréter comme correspondant à une légère consolidation du matériau sous l'effet de décharges locales. Nous proposons de donner à ce phénomène le nom de consolidation post-fluage.

5. Une application intéressante concerne la détermination du coefficient K de BAILEY-NORTON au moyen de l'essai de traction-relaxation proposé par J. LEMAITRE. Au premier instant de la relaxation déclenchée quand $\varepsilon = \varepsilon_\rho$, θ passe instantanément de la valeur $\theta(\varepsilon_\rho | \dot{\varepsilon}) = \theta_\rho$ atteinte en fin de traction à la valeur $\theta(\varepsilon_\rho | 0) = \theta_{c\rho}$ comprise entre θ_ρ et 1. Cette valeur $\theta_{c\rho}$ figure dans la formule donnant ensuite la variation de la contrainte [2] :

$$k = (1 - \theta_{c\rho})\varepsilon_\rho + K\, \theta_{o\rho}^{1-1/n} (\dot{\varepsilon}^\rho)^{1/n} \qquad (2)$$

Par adaptation optimale de (2) à la courbe expérimentale k(t), on détermine n et une première valeur de K, soit K_R. La relation (1) fournissant d'autre part λ , on en déduit, pour la même valeur de n, une seconde valeur de K, soit K_T. Ces deux valeurs doivent être égales aux erreurs d'expérience près. En général, elles diffèrent notablement si l'on prend $\theta_{o\rho} = \theta_\rho = 1$, c'est-à-dire si l'on néglige la consolidation.

Exemples - 1°) Cuivre du commerce de pureté 99,96% recuit à 600°C pendant 4 Heures. Vitesse d'étirement : 0,72. 10^{-3} sec^{-1}. Température de l'essai : 20°C.

Sans consolidation : $\theta_{o\rho} = 1$, n = 101,9 $K_R.10^4 = 22{,}01$ $K_T.10^4 = 8{,}68$

Avec consolidation : $\theta_{o\rho} = 0{,}991$ n = 28,29 $K_R.10^4 = 10{,}50$ $K_T.10^4 = 10{,}44$

2°) Acier Z2CN 18-10 ayant subi un recuit d'austénisation à 1050° pendant une heure, puis une trempe à l'eau. Vitesse d'étirement : 0,97.10^{-5} sec^{-1}. Température de l'essai : 20°C.

Sans consolidation : $\theta_{o\rho} = 1$ n=55,83 $K_R.10^4 = 15{,}89$ $K_T.10^4 = 12{,}82$

Avec consolidation $\theta_\rho = 0{,}988$ $K_R.10^4 = K_T.10^4 = 13{,}55$ pour $\theta_{o\rho} = 0{,}993$ compris entre 0 et 1 (n : 43,92).

6. Lorsqu'on tient compte de la consolidation, un accord satisfaisant entre K_R et K_T pour la même valeur de n est l'indice de

la stabilité du matériau vis à vis de $\dot{\varepsilon}$. Il est intéressant de constater qu'il existe une corrélation entre cette stabilité (n et K constants $\forall \dot{\varepsilon}$) et la stabilité structurale du matériau. Dans le second exemple cité, l'acier Z2CN18-10 apparaît comme stable à la vitesse extrêmement faible de l'expérience alors que sa structure austénitique est instable à 20°C. Ce paradoxe apparent s'explique par la transformation, dès l'entrée en plasticité, de l'austénite instable en martensite stable en quantité d'autant plus importante que la vitesse d'étirement est plus faible [3] , c'est-à-dire ici en quasi totalité.

Inversement, un désaccord entre K_R et K_T serait l'indice d'une instabilité structurale comme le montrerait une étude systématique du même acier à différentes vitesses de traction. En effet, aux vitesses supérieures à celle de l'exemple cité, la quantité de martensite formée diminue et K devient, à n constant, une fonction décroissante de $\dot{\varepsilon}$.

BIBLIOGRAPHIE

[1] I.U.T.A.M. Symposium EAST-KILBRIDE , 1968, Springer, Wien, 1979, p. 170.

[2] Comptes-Rendus de l'Académie des Sciences, 284, Série B, GAUTHIER-VILLARS, Paris, 1977, p. 9.

[3] ROUSSEAU et al. Revue de Métallurgie. Mém. Scientifiq. LXVII, 1970, p. 315.

The Use of an Empirical Yield Surface Model in Predicting Shakedown Loads

D. N. MORETON and D. G. MOFFAT

Department of Mechanical Engineering
University of Liverpool
P.O. Box 147
Liverpool, England, L69 3BX.

1. Introduction

The Bleich/Melan Theorem permits shakedown loads to be calculated where the assumption of elastic/perfectly plastic material idealisation is realistic. For materials which demonstrate significant strain hardening characteristics an arbitrarily chosen proof-stress is often used in conjunction with this theorem. However, this technique will usually lead to an over-estimate of the shakedown load and the need exists for a more soundly based method of prediction. Any such method will require a re-statement of the Bleich/Melan Theorem to include strain-hardening effects. One such re-statement has been given together with a proof by Kraus and Rice (1) although this proof places considerable restrictions on the yield surface movements permitted. However, in this present work the Kraus and Rice Theorem is used together with an empirical yield surface model, to predict the shakedown pressures of three pressure vessel components manufactured from a strain-hardening material (stainless steel type 321).

2. Statement of the Shakedown Theorem (Kraus and Rice)

"If any time independent distribution of residual stresses can be found such that the sum of these residual stresses and the perfectly elastic stresses is a safe stress with respect to the yield surface distribution at any one time in the loading history, the body will shakedown".

3. An Empirical Model to Describe the Yield Surface Movement

The model used here was derived by the authors (see (2) and (3)) for a range of steels applicable to the pressure vessel industry. It is simply stated here for the two dimensional principal stress plane as

$$C = g(\gamma).\sigma_{PE} - g(\gamma).\sigma_y^*$$

where C = Length of stress vector from stress origin to the translated yield locus centre

σ_{PE} = Magnitude of the effective pre-stress

σ_y^* = Yield stress of the virgin material in tension

and
$$g(\gamma) = \frac{1}{1.115}\left[\frac{1}{1 - \frac{1}{2}\sin 2\gamma}\right]^{\frac{1}{2}}$$

where γ defines the pre-stress direction (measured +ve anticlock from the +ve abscissa). This notation is illustrated in Fig. 1.

4. Elastic-Plastic Stress Path on the σ_1 v σ_2 Plane

In order to predict the shakedown load for a structure, a knowledge of the state of stress with increasing load for the first yielding location of the structure is required. This could be obtained from an elastic/plastic numerical analysis of the structure. However, for the examples presented here, experimentally obtained strains were available, and the computer program (4) was used to calculate the stresses.

5. Evaluation of the Shakedown Load - Experimental and Theoretical

Three pressure vessel components, two torispherical drumheads and a nozzle in sphere junction, were manufactured from stainless steel type 321, extensively strain gauged and pressure cycled to find the maximum shakedown pressure (see (5)). Tensile specimens of this material were used to obtain the yield stress.

With a knowledge of the pressure/strain performance of the first yielding location of each component, the principal stress path was computed (using (4)). Using the yield surface model at each increment of pressure the current yield surface was established. Assuming an elastic stress path during unloading the residual stresses were obtained. Where these lay within the current yield surface shakedown was predicted and where outside - no shakedown predicted.

Fig. 2 shows this procedure for one of the drumheads, where at a pressure of 310 lbf/in^2 (2.137 MN/m^2), the residual stresses lie just on the current yield locus. The experimental shakedown pressure for this vessel was found to be 340 lbf/in^2 (2.344 MN/m^2). Table 1 gives experimental and predicted values of shakedown pressure for the three vessels using this technique.

Vessel Designation	P_S MN/m^2 (lbf/in^2)		% Error
	Exp.	Predicted	
Head 32	2.344 (340)	2.137 (310)	8.8
Head 31	1.62 (235)	1.462 (212)	9.8
Nozzle	2.41 (350)	1.958 (284)	18.8

TABLE 1

6. Conclusions from the Work to Date

Using the shakedown theorem of Kraus and Rice with the yield surface model of Moreton et al. ((2) and (3)) has given shakedown pressures for three stainless steel pressure vessel components with acceptable accuracy. It should be noted that the empirical model of (2) and (3) was formulated using different materials to that used in the manufacture of the test vessels (5). It is thus implied that the model will give acceptable shakedown predictions for other pressure vessel steels.

7. References

1. Kraus, H.; Rice, J.S.: Concerning shakedown of a strain hardening continuum. Rensselaer Polytechnic Institute, Hartford Graduate Centre, East Windsor Hill, Connecticut, Internal Report (March 1970).

2. Moreton, D.N.; Moffat, D.G.; Hornby, R.P.: Techniques for investigating the yield surface behaviour of pressure vessel materials. J. of Strain Analysis, Vol 13, No 3 (1978).

3. Moreton, D.N.; Moffat, D.G.; Parkinson, D.B.: The yield surface behaviour of pressure vessel steels. To be submitted to the J. of Strain Analysis.

4. Moreton, D.N.; Moffat, D.G.: A computer program to calculate post yield stresses from strains measured on strain hardening materials. Strain, J. of Brit. Socy. for Strain Measurement, Vol 11, No 2 (1975).

5. Moreton, D.N.; Moffat, D.G.: Shakedown of three stainless steel pressure vessel components. 3rd Int. Conf. on P.V. Tech., Tokyo (1977) 233-245.

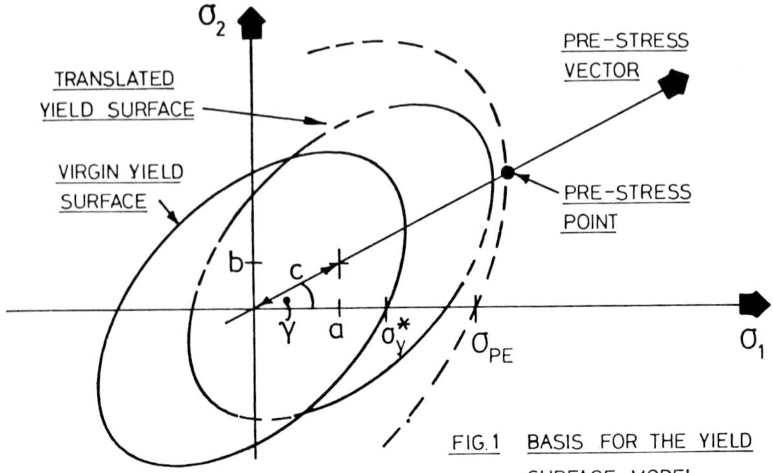

FIG. 1 BASIS FOR THE YIELD SURFACE MODEL

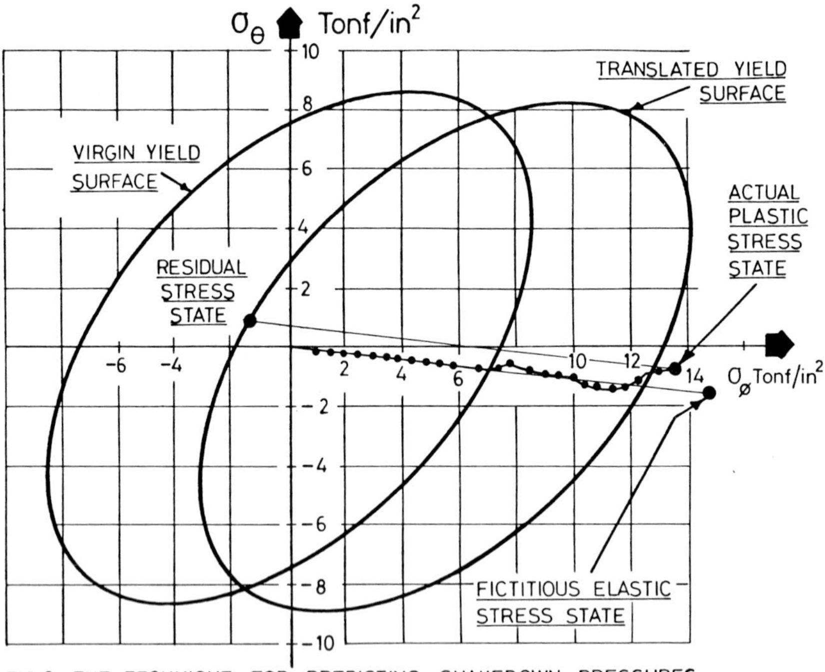

FIG. 2 THE TECHNIQUE FOR PREDICTING SHAKEDOWN PRESSURES.

Plasticity Solutions by Finite Element Method

G.C.NAYAK

Professor of Civil Engineering
University of Roorkee
Roorkee 247672 INDIA

Summary

A general non-linear formulation for materially non-linear problems using local derivatives of shape functions is presented. The stable 'Alpha Constant' stiffness method with a new two step process is also developed as an economical alternative for the solution of problems with physical non-linearities.

Introduction

A continuous effort is still going on to develop computer programs which contain general plasticity laws, family of elements, variety of solution algorithms and systematic convergence tests. The finite element literature dealing with the subject is now large and differences lie both in the apparent forms of solution algorithm and in the form of constitutive relations postulated. In all processes a linearisation is used to obtain a solution iteratively. In some, however, the stiffness matrix is continuously updated while in others a constant linear matrix is used requiring a single matrix inversion. The introduction of residual force concept is an essential feature of any rational approach to non-linear problems. The Alpha constant stiffness method has now turned out to be a very versatile, convergent and economical method and one can solve problems of strain softening, non-associated and cyclic load problems without any difficulty. This method has been extended to several non-linear field problems.

A unified formulation developed by the author is based on the concept of generalised modulus and generalised stresses. The Alpha-constant stiffness method has been modified for two increments at a time to economise the computational effort and is currently under testing for three steps.

The author has introduced the isoparametric elements for the solution of non-linear problems by using numerical integration technique. Once a smooth distribution of displacements across the element postulated by a shape function is defined, the state of strain and stress at each integrating point can be examined individually and the resulting element properties are obtained by a suitable weighted summation over the integrating points. It may be pointed out that the region around the each integrating point (usually known as the Gauss point) acts as a sub-element of a prescribed higher order element. This has a close relationship with mesh refinement and has also been used for economising iterative solutions by using coarse mesh solutions for a corresponding fine mesh sub-divisions.

A General Formulation

The three basic evaluations in a plasticity program are(1)

1) Element stiffness given by

$$[K_T] = \int_v [B]^T [D_{ep}] [B] dV \tag{1}$$

2) Evaluation of residuals

$$\{\Psi\} = \{R\} - \int_v [B]^T \{\sigma\} dV \quad \text{and} \tag{2}$$

3) The evaluation of equivalent nodal forces due to various actions.

It is well known that the $[B]$ matrix consists of global derivatives of the shape functions given by

$$[B] = [L] [N] \tag{3}$$

where operator $[L]$ is a differential operator of global derivatives. The global operator $[L]$ can be written as

$$[L] = [m] [\ell] \tag{4}$$

where for a two dimensional problem the matrix

$$[L] = \begin{bmatrix} \frac{\partial}{\partial x} & 0 \\ 0 & \frac{\partial}{\partial y} \\ \frac{\partial}{\partial y} & \frac{\partial}{\partial x} \end{bmatrix}_{3 \times 2} = \underset{3x4}{[m]} \begin{bmatrix} I_2 & \frac{\partial}{\partial \xi} \\ I_2 & \frac{\partial}{\partial \eta} \end{bmatrix}_{4 \times 2}$$

where $I_2 = \begin{bmatrix} 1 & 0 \\ 0 & 1 \end{bmatrix}$ and $[m] = [L(\xi), L(\eta)]$, and for a 3D problem

$$[m] = [L(\xi), L(\eta), L(\zeta)] \tag{5}$$

Thus Eq.(3) and (4) give

$$[B] = [m][\ell][N] = [m][g] \tag{6}$$

with $[g] = [\ell][N]$

Substituting Eq.(6) into Eq.(1), we get

$$[K_T] = \int_V [g]^T [D_g] [g] \, dV \tag{7}$$

with the generalised modulus matrix

$$[D_g] = [m]^T [D_{ep}] [m] \tag{7a}$$

and

$$[\psi] = \{R\} - \int_V [g]^T \{\sigma^*\} \, dV \tag{8}$$

with the generalised stresses

$$\{\sigma^*\} = [m]^T \{\sigma\} \tag{9}$$

Eqs.(7) and (9) clearly require recalculation of $[D_g]$ and $\{\sigma^*\}$ and the larger matrices $[g]$ are calculated only once for one parent element in the form of local derivatives of shape functions $[N]$.

For plasticity solutions the formulations of Ref.(1) gives several useful forms of $[D_g]$ given by Eq.(7a) and $\{\sigma^*\}$ given by Eq.(9). These can be utilised for developing various alternative solution algorithms.

Two Step Alpha Constant Stiffness Method

A variety of solution algorithms including Alpha Constant Stiffness method for the solution of plasticity problems are discussed by Nayak and Zienkiewicz(1). The modification of Alpha method(4) gives convergent solutions. Summarising the single step method(1-4), we have

$$\left[K_T\right]\Delta\{\delta\}^i = \left[K_o + K_{nl}\right]\Delta\{\delta\}^i = \{\psi\}_i \quad \text{and} \tag{10}$$

$$\left[K_o\right]^{-1}\{\psi\}_i = \Delta\{\delta_o\}^i = \{\alpha_i\}\Delta\{\delta_o\}^i \tag{11}$$

The residual vector as derived in Ref.(4)

$$\{\psi\}_{i+1} = \left[K_o\right](I-\alpha_i)\Delta\{\delta_o\}^i - \left[K_{nl}\right]\left[\alpha_i\right]\Delta\{\delta_o\}^i \tag{12}$$

and

$$\Delta\{\delta_o\}^i = \left[K_o\right]^{-1}\{\psi\}_{i+1}$$

$$= (I - \alpha_i)\Delta\{\delta_o\}^i + \left[K_o\right]^{-1}\{\psi\}_{i+1} \tag{13}$$

where $\{\bar{\psi}\}_{i+1} = -\left[K_{nl}\right]\left[\alpha_i\right]\Delta\{\delta_o\}^i$ (14)

For two step solutions two load increments, $\Delta\{P\}_1$ and $\Delta\{P\}_2$ are applied simultaneously on constant stiffness $\left[K_o\right]$ and two sets of diagonal matrix $\left[\alpha_1\right]$ and $\left[\alpha_2\right]$ are obtained such that

$$\left[K_o + K_{nl}\right]\Delta\{\delta\}_1 = \Delta\{P\}_1 = \left[K_o\right]\Delta\{\delta_o\}_1$$

and

$$\left[K_o + K_{nl}\right]\Delta\{\delta\}_2 = \Delta\{P\}_2 = \left[K_o\right]\Delta\{\delta_o\}_2$$

so that

$$\Delta\{\delta\}_1 = \left[\alpha_1\right]\Delta\{\delta_o\}_1 \quad \text{and} \quad \Delta\{\delta\}_2 = \left[\alpha_2\right]\Delta\{\delta_o\}_2 \tag{15}$$

Several alternatives for evaluation of $\left[\alpha_1\right]$ and $\left[\alpha_2\right]$ as given in Eqs. (10 to 14) exist. Extrapolation techniques depending upon the nonlinearity of the problem are also possible. The simultaneous solution of two right hand sides and evaluation of $\left[\alpha_1\right]$ and $\left[\alpha_2\right]$ give further reduction in computational costs of non-linear solutions.

Conclusions

Several advantages of unified formulation based upon concept of generalised modulus and generalised stresses are obtained. The organisation of computer programs becomes simpler and the number of arithmetic operations for 3D problems(3) are reduced considerably. The advantage of simultaneous solution with two right hand sides and two diagonal $[\alpha]$ matrix give stable two step Alpha method giving further computational efficiency over Newton-Raphson method.

Several problems of pressure vessel with different head shapes, nozzle junction, notched beam specimen, extrusion and polystyrene crazing have been solved with the aforesaid technique.

References

1 Nayak, G.C.: Zienkiewicz, O.C., 'Elasto-Plastic Stress Analysis - A Generalisation for Various Constitutive Relations including Strain Softening', IJNME, 5, 113-135, 1972.

2 Nayak, G.C.: Zienkiewicz, O.C., 'Note on the 'Alpha' Constant - Stiffness Method for the Analysis of Non-linear Problems', IJNME, 4, 579-582, 1972.

3 Nayak, G.C., 'Plasticity and Large Deformation Problems by the Finite Element Methods', Ph.D. Thesis, University of Wales, Swansea, 1971.

4 Cochet, J.F. :Dhatt, G.S., 'A Modification of the Alpha Constant Technique', IJNME, 11, 1053-1054, 1977.

About the Periodic Contact Problem for the Maxwell Body

Bernard NAYROLES

Laboratoire de Mécanique et d'Acoustique - C.N.R.S.
31, chemin Joseph Aiguier 13274 Marseille Cédex 2.

Summary

This paper is devoted to the periodic problem for a Maxwell body submitted to unilateral constraints. A similar subject was already treated by anthor's group ([1],[2],[3]) about periodic stresses in cracked turbine blades. The unilateral constraint was the non-negativity of the crack opening displacement.

The new results brought by this paper are the complete formulations of the periodic problem for any convex domain of unilateral constraints as well as a new technique which allows to reach the solution directly, that is to say with no regularization of the constitutive equation. Due to this technique that exactly fits the liquid character of the Maxwell body, the shape of the obstacle can be as irregular as wanted. It is hoped that the tools used here for the case of a beam can be generalized to other cases.

§ 1. Naīve equations

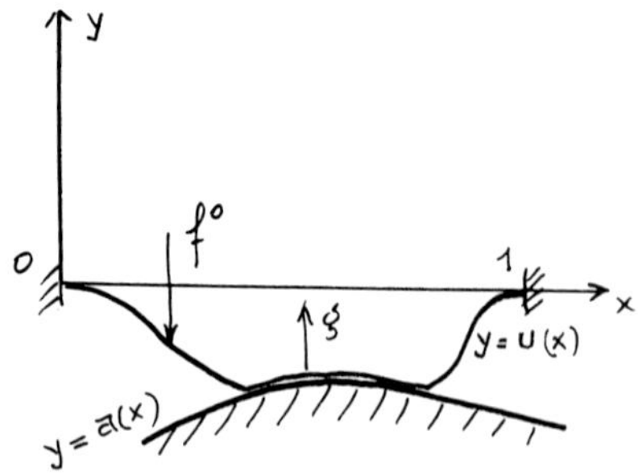

Consider a straight beam, with unity length, flexured in the (x, O, y) plane as represented by the opposite figure. According to the simplest flexion theory the deflection u(x) is small and parallel to the y-axis. The generalized strain e is the curvature $\frac{\partial^2 u}{\partial x^2}$, associated with the generalized stress s which is the bending moment.

This beam is fixed at both ends and submitted to a given imposed strain $-e^\circ$, and to a given lineic density of forces f° ; both of them depend on (x, t). Moreover a fixed obstacle is given by its equation y = a(x) and the beam must remain above it. A suitable generality is obtained by allowing the function a to take the value $-\infty$.

The quasi-static evolution of such a beam is ruled by the following system :

$$u(0) = u(1) = \frac{\partial u}{\partial x}(0) = \frac{\partial u}{\partial x}(1) = 0 \tag{1}$$

$$e = \frac{\partial^2 u}{\partial x^2} + e^\circ \tag{2}$$

$$\frac{\partial^2 s}{\partial x^2} = f^\circ + g \tag{3}$$

(g lineic density of constraint forces associated with the obstacle)

$$\left.\begin{array}{l} u \geqslant a \\ g \geqslant 0 \\ ug = 0 \end{array}\right\} \tag{4}$$

$$s(x,t) = \mathcal{K}(x)\,(e(x))\,(t) \tag{5}$$

The last equation describes the behaviour of the material : it is a functional relation with respect to t, pointwise with respect to x. To shorten the exposal let us restrict ourselves to the Maxwell body, as an example.

Its equations are :

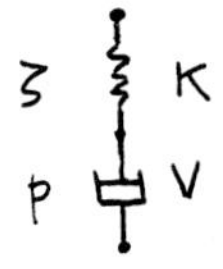

$$\begin{array}{l} e = \zeta + p \\ s = K\zeta = V\,\dot{p} \end{array} \tag{6}$$

as represented by the opposite rheological model.

It is supposed that both K(x) and V(x) are uniformly upper and lower bounded on] 0, 1 [.

§ 2. Periodic constitutive equations and functional framework

Let us suppose that $\dot{e}^\circ$ and f° are T-periodic time functions whereas K and V are time independant. Suppose that s is also T-periodic and put :

$$\bar{s} = \frac{1}{T}\int_0^T s(t)\,dt \quad , \hat{s} = s - \bar{s} \tag{7}$$

Then equations (6) can be reduced to their "periodic form" :

$$\left.\begin{aligned} e(t) &= \alpha t + e_1 + \hat{e}(t) \\ \alpha &= V^{-1}\bar{s} \\ \hat{e}(t) &= K^{-1}\hat{s}(t) + \int_0^t V^{-1}\hat{s}(\theta)d\theta - \frac{1}{T}\int_0^T dt\int_0^t V^{-1}\hat{s}(\theta)\,d\theta \end{aligned}\right\} \tag{8}$$

and the constant term e_1 is arbitrary.

According to (8) and (2) the deflection u takes the form :

$$u(t) = \xi t + u_1 + \hat{u}(t)$$

and the specificity of the situation lays in the presence of the secular term ξt as well as in the arbitraryness of the constant term u_1 with respect to the constitutive equation.

We now introduce the shortened notations :

$$E = L^2\,(]0,1[),\; U = H_0^2\,(]0,1[),\quad \Phi = H^{-2}\,(]0,1[)$$

and , for X denoting any of this three spaces,

$$\tilde{X} = L^2\,(]0,T[, X)$$

is isomorphic to the space of T-periodic square-summable time functions with values in X. The hilbert space $\tilde{X}$ admits the hibertian suuplementary subspaces :

$$\tilde{X} = \bar{X} + \hat{X}$$

where $\bar{X}$, isomorphic to X, is the space of constant functions, $\hat{X}$ the space of functions with mean value 0. We'll use the following notations for scalar or duality products :

$$(u(t), f(t)) \in U \times \Phi \mapsto u(t).f(t) = \int_0^1 u(x,t)\,f(x,t)\,dx$$

$$(u,f) \in \tilde{U} \times \tilde{\Phi} \to \langle u,f\rangle = \int_0^T u(t).f(t)\,dt = T\,\bar{u}.\bar{f} + \langle\hat{u},\hat{f}\rangle$$

$$(e(t),s(t)) \in E \times E \mapsto e(t).s(t) = \int_0^1 e(x,t)\,s(x,t)\,dx$$

$$(e,s) \in \tilde{E} \times \tilde{E} \mapsto \langle e,s\rangle = \int_0^T e(t).s(t)dt = T\,\bar{e}.\bar{s} + \langle\hat{e},\hat{s}\rangle$$

The functional choice adopted here is :

$$\left.\begin{array}{l} \alpha \in E \ ; \ \xi \in U \\ \hat{e} \in \hat{E} \ ; \ \hat{u} \in \hat{U} \\ \dot{e}^{\circ},\ s \in \tilde{E};\ f^{\circ},\ g \in \tilde{\Phi} \end{array}\right\} \qquad (9)$$

and no regularity condition is required for u_1. Equations (8) yield :

$$\exists m > 0 \quad < \hat{e}, \hat{s} > \ = \ < K^{-1}\hat{s}, \hat{s} > \ \geqslant m \quad \|\hat{s}\|^2_{\tilde{E}} \qquad (10)$$

The product Hilbert space $E \times \hat{E}$ is equipped with the scalar product :

$$((\alpha,\hat{e}),\ (\bar{s},\hat{s})) \mapsto T^2 \ \alpha.\bar{s} + < \hat{e}, \hat{s} > \ = \ <<(\alpha,\hat{e}),(\bar{s},\hat{s})>>$$

with respect to which the periodic form (8) of constitutive equations comes to a continuous and coercive linear mapping $\mathcal{K}_p^{-1}$

$$(\bar{s}, \hat{s}) \mapsto (\alpha, \hat{e}) = \mathcal{K}_p^{-1} (\bar{s}, \hat{s}) \qquad (11)$$

as easily derived from (8) and (10).

Moreover when (11) holds one may write :

$$\| \hat{e}(t+\Delta t) - \hat{e}(t) \|_E \leqslant M_1 \quad \hat{s}\ (t+ \Delta t) - \hat{s}(t) \|_E + M_2\ \Delta t\ \|s\|_{\tilde{E}}$$

for suitable constants M_1 and M_2. (12)

§ 3. On unilateral constraints

The complementarity system (4) must be precised in terms of the previous functional spaces. First of all remark that a Maxwell body is a liquid is such a way that we must expect for u_1 the same irregularities as those possessed by a ; for instance, if f° is reduced to the gravity forces only, an obvious periodic solution is the rest of the beam on the obstacle and a permanent flow out of it. Hence u_1 may be taken equal to a.

To escape a part of these difficulties it is convenient to replace the unknown u_1 by :

$$b = a - u_1$$

which may be expected a bit more regular, but not nessarily in U.

Let us define in U the set :

$$\mathbf{C}(b) = \{u \in U / \forall x\ ]0,1[\quad u(x) \geqslant b(x)\} \qquad (13)$$

which is a closed convex set, as easily checked. It is classical to replace (4) by the variational inequation :

$$\xi t + \hat{u}(t) \in \mathbf{C}(b) \;,\quad \forall v \in \mathbf{C}(b) : [v-\xi t-\hat{u}(t)].g(t) \geqslant 0 \tag{14}$$

This is a less intuitive formulation than given by the following :

Proposition 1. Assume that b is the pointwise infimum of some family of U :

$$b(x) = \inf_i v_i(x) \;,\quad v_i \in U$$

Then (14) is equivalent to :

$$u \in \mathbf{C}(b),\ g \geqslant 0,\ \text{supp}\,(g) \subset \{x \in \,]0,1[\ /\ u(x)=b(x)\} \tag{15}$$

where the inequality is taken in the sense of positive distributions.

Let us now introduce s°, some particular solution of

$$\frac{\partial^2 s^\circ}{\partial x^2} = f^\circ$$

$$\sigma = s - s^\circ \quad , \quad \varepsilon = e - e^\circ, \quad \gamma = \alpha - \dot{\bar{e}}^\circ$$

Let us also introduce

$$\Gamma(b) = \{e \in E\ /\ \exists\ u \in \mathbf{C}(b) : e = \frac{\partial^2 u}{\partial x^2}\ \}$$

It is fairly classical (sec [1], [2]) that the naïve system initially written comes down to :

$$\gamma t + \hat{\varepsilon}(t) \in \Gamma(b) \;, \forall \eta \in \Gamma(b) : \quad [\eta - \gamma t - \hat{\varepsilon}(t)]\ .\sigma(t) \geqslant 0 \tag{16}$$

$$(\gamma, \hat{\varepsilon}) = \mathcal{K}_\rho^{-1}\ (\bar{\sigma}, \hat{\sigma}) + \mathcal{K}_\rho^{-1}\ (\dot{\bar{s}}^\circ, \hat{s}^\circ) - (\dot{\bar{e}}^\circ, \hat{e}^\circ) \tag{17}$$

which will be now studied. Notice that (16) is nothing but the expression of (14) in terms of stresses and strains.

§ 4. A new formulation of the periodic problem

The presence of the secular terms ξt, or γt, makes hopeless the research of a solution of $\{(16),(17)\}$ for every $t \in\ ]-\infty, +\infty[$. Indeed the secular term $\xi(x)t$ should be zero at those points x where a(x) is finite. Easy counter examples may be given of the impossibilité of any solution, for instance in the case when f° is reduced to an upward gravity.

If one keeps in mind that a periodic solution is mainly interesting because of it asymptotic character for large positive t, one is led to give the "time one-sided" formulation :

Problem I. Find $(\gamma,\hat{\varepsilon},\sigma, b, t_\circ)$ solution of (16) and (17) for

$t \geqslant t_\circ$

A fairly easy proof may be given for the equivalence of problem I to

Problem II. Find $(\gamma,\hat{\varepsilon},\sigma, b)$ solution of (17) and of

(18) $\gamma \in \mathbf{K} \quad \forall \gamma' \in \mathbf{K} : \quad (\gamma'-\gamma).\bar{\sigma} \geqslant 0$

(19) $\forall t \in]0,T[\quad \hat{\varepsilon}(t) \in \Gamma(b), \quad \forall \varepsilon' \in \Gamma(b) : [\varepsilon'-\hat{\varepsilon}(t)].\sigma \geqslant 0$

In (18) $\mathbf{K}$ denotes the asymptotical cone of $\Gamma(b)$, defined by :

$$\mathbf{K} = \{\gamma \in E / \forall \lambda \in \mathbb{R}^+, \forall \varepsilon_\circ \in \Gamma(b) \quad : \quad \varepsilon_\circ + \lambda\gamma \in \Gamma(b)\}$$

It can be seen that $\mathbf{K}$ does not depend on u_1 but only on the "effective domain" of a ; more precisely $\mathbf{K}$ is the image :

$$\mathbf{K} = \frac{\partial^2}{\partial x^2} \quad (\{\xi \in U / \ a(x) > -\infty \Rightarrow \xi(x) \geqslant 0\} \ )$$

and the set between parenthesis is the asymptotical cone of $\mathbf{C}(b)$.

Let us now put :

$$\tilde{\Gamma}(b) = \{ e \in \tilde{E} / \ \text{a.e.}\ t \in]0,T[\quad : \varepsilon(t) \in \Gamma(b)\}$$

and introduce the indicator functions $\Psi_{\Gamma(b)}$, $\Psi_{\tilde{\Gamma}(b)}$ and the support functions $\Psi^*_{\Gamma(b)}$, $\Psi^*_{\tilde{\Gamma}(b)}$. As it is well known the variational inequality :

$$\text{a.e.}\ t \in \ ]0, T[\quad : \quad \varepsilon(t) \in \Gamma(b), \forall \varepsilon' \in \Gamma(b) \ : [\varepsilon'-\varepsilon(t)].\sigma \geqslant 0 \tag{20}$$

is equivalent to any of the three next ones :

$$\hat{\varepsilon} \in \tilde{\Gamma}(b), \ \forall \varepsilon' \in \tilde{\Gamma}(b) \ : \ \langle \varepsilon'-\hat{\varepsilon} \ ,\sigma\rangle \ \geqslant 0 \tag{21}$$

$$\Psi_{\tilde{\Gamma}(b)}(\hat{\varepsilon}) + \Psi^*_{\tilde{\Gamma}(b)}(-\sigma) + \langle\hat{\varepsilon} \ \ \sigma\rangle = 0 \tag{22}$$

$$\forall \tau \in \tilde{E} \ : \langle\hat{\varepsilon}, \ \tau-\sigma\rangle + \Psi^*_{\tilde{\Gamma}(b)}(-\tau) - \Psi^*_{\tilde{\Gamma}(b)}(-\sigma) \geqslant 0 \tag{23}$$

A main result is now :

Proposition 2. Assume that b is the pointwise infimum of some family $\{v_i\}$ in U, and lower-bounded. Then :

$$g = \frac{\partial^2}{\partial \ z} \geqslant 0 \Rightarrow \ \Psi^*_{\tilde{\Gamma}(b)}(-\sigma) = \Psi^*_{\tilde{\mathbf{C}}(b)}(-g) = T\Psi_{\mathbf{C}(b)}(-\bar{g})$$

$$g = \frac{\partial^2\sigma}{\partial x^2} \not\geqslant 0 \Rightarrow \ \Psi^*_{\tilde{\Gamma}(b)} = +\infty$$

In this statement $\tilde{\mathbf{C}}(b)$ is $\{u \in \tilde{U} / \ \text{a.e.}\ t \in]0,T[\ : \ u(t) \in \mathbf{C}(b)\}$

The Lebesque theorem involves that :

$$\int_0^T b.\ g(t)dt = \inf \int_0^T v_i.\ g(t)dt = \inf T\ v_i.\bar{g} = -T\Psi^*_{\mathbf{C}(b)}(-\bar{g})$$

which is the main point of proposition 2. Others are obvious. Now inequation (23) may be restricted to those τ whose mean value is preciselu $\bar{\sigma}$; then $\hat{\sigma}$ satisfies :

$$\hat{\sigma} \in \Lambda(\bar{\sigma}), \forall \hat{\tau} \in \Lambda(\bar{\sigma}) \quad : \langle \hat{e}, \hat{\tau} - \hat{\sigma} \rangle \geqslant 0 \tag{24}$$

with
$$\Lambda(\bar{\sigma}) = \{ \tau \in \hat{E} / \frac{\partial^2}{\partial x^2} (\bar{\sigma} + \tau) \geqslant 0 \}$$

which is a convex closed set in $\hat{E}$. Now (18) is equivalent to

$$-\bar{\sigma} \in K^\circ \quad \forall \bar{\tau} \in -K^\circ \quad T^2\gamma \,.\, (\bar{\tau} - \bar{\sigma}) \geqslant 0 \tag{25}$$

and $\{(24), (25)\}$ is equivalent to the single inequality :

$$(\bar{\sigma}, \hat{\sigma}) \in \mathbf{P}, \forall (\bar{\tau}, \hat{\tau}) \in \mathbf{P} : T^2\gamma . (\bar{\tau} - \bar{\sigma}) + \langle \hat{\varepsilon}, \hat{\tau} - \hat{\sigma} \rangle \geqslant 0 \tag{26}$$

with : $\mathbf{P} = \{ (\bar{\sigma}, \hat{\sigma}) / \bar{\sigma} \in -K^\circ, \hat{\sigma} \in \Lambda(\bar{\sigma}) \}$, non empty closed and convex set.

Due to the continuity and the coercivity of $\mathcal{K}_\varphi^{-1}$ system $\{(17), (26)\}$ admits a unique solution $(\gamma, \hat{\varepsilon}, \bar{\sigma}, \hat{\sigma})$.

The resolution of problem II will be achieved if we find b such that (19) actually holds, as well as the asumptions done about b.

<u>Theorem</u> Assume that e° and f° belong to $H^1(]0,T[,E)$, $H^1(]0,T[,\Phi)$ respectively. Then problem II admits solutions $(\gamma, \hat{\varepsilon}, \bar{\sigma}, \hat{\sigma}, b)$ with a unique $(\gamma, \hat{\varepsilon}, \bar{\sigma}, \hat{\sigma})$ solution of system $\{(17),(26)\}$ and

$$\hat{\varepsilon}, \hat{\sigma} \in H^1(]0, T[, E) \tag{27}$$

The function b may be choose in any way provided that :

$$\forall x \in \text{supp}\left(\frac{\partial^2\sigma}{\partial \chi^2}\right) : \quad b(x) = \min_{t \in [0,T]} \hat{u}(x,t)$$

$$\forall x \in]0,1[\quad : b(x) \leqslant \min_{t \quad [0,T]} \hat{u}(x,t)$$

where $\hat{u}$ obviously denotes the unique element of $H^1(]0, T[, U)$ such that $\frac{\partial^2 \hat{u}}{\partial x^2} = \hat{e}$

A first step of the proof is to establish the continuity of $\hat{\varepsilon}$ and $\hat{\sigma}$. From (26) one can derive, for any positive Δt :

$$\langle \hat{\varepsilon}, \hat{\sigma}(.+\Delta t) - \hat{\sigma} \rangle \geqslant 0$$

and
$$\langle \hat{\varepsilon}(.+\Delta t), \hat{\sigma} - \hat{\sigma}(.+\Delta t) \rangle \geqslant 0$$

which yield :

$$< \hat{\varepsilon}\ (.+\Delta t) - \hat{\varepsilon},\ \hat{\sigma}(.+\Delta t) - \hat{\sigma} > \ \leqslant 0$$

Then (8), (10) and (12) and usual regularization arguments provide the regularity defined by (27)

Notice that $\hat{u} \in H^1$ ( ]0,T[, H^2_o]0, 1[ ) implies that

$$(x,t) \mapsto \hat{u}(x,t)$$

is continuous ; hence we may write

$$b(x) = \min_{t \in [0,T]} \hat{u}(x,t)$$

and, moreover, b is lower-bounded. Hence it satisfies all previous assumptions.

Now the duality properties yield that $g = \frac{\partial^2 u}{\partial x^2}$ satisfies :

$$g \geqslant 0, \quad \forall h \geqslant 0, \ \bar{h} = \bar{g} \ : \ <\hat{u},\ h-g> \ \geqslant \ 0 \qquad (28)$$

Lemma. Provided $(s,t) \mapsto \hat{u}(s,t)$ is continuous (28) is equivalent to :

$$g \geqslant 0,\ \text{supp.}(g) \subset \{(x,t) \in\ ]0,1[\ \times [0,T] / \hat{u}(x,t) = \min_{\theta \in [0,T]} u(x,\theta)\} \qquad (29)$$

Indeed (29) obviously involves (28). Conversely suppose :

$$\hat{u}\ (x_o,t_o) \ > b(x_o) + \varepsilon\ ,\ (x_o,\ t_o) \in \text{supp. } g$$

There exist a positive number r and a time t_1 such that

$$\hat{u}(x,t) > b(x_o) + \varepsilon \quad \text{if } \sqrt{(x-x_o)^2 + (t-t_o)^2} \leqslant r$$

$$\hat{u}(x,t) < b(x_o) + \varepsilon \quad \text{if } \sqrt{(x-x_o) + (t-t_1)^2} \leqslant r$$

Let ϕ be a non-negative test-fonction, smaller than 1, with support embedded in the ball of center $(x_o,\ t_o)$ and radius r.

Put :

$$h = g - g\phi \ + (g\phi)\ (.,.+t_o-t_1)$$

It is easy to check that h is a non negative distribution, with the same mean value as g and

$$<\hat{u}\ ,\ h-g> \ < \ 0$$

which is contrary to (28). Hence (28) implies (29).

Now, exactly as when proposition 1 was concerned, relation (29) implies, as $\hat{u}$ belongs to $\tilde{\mathbf{C}}(b)$:

$$\forall v \in \tilde{\mathbf{C}}(b) \qquad <v-\hat{u},\ g> \ \geqslant 0$$

which yields that (21), (20) and also (19) hold.

References :

[1] R. Bouc, G. Geymonat, M. Jean, B. Nayroles.
Problèmes unilatéraux hilbertiens pour les corps viscoélastiques fissurés.
Note L.M.A. n° 1420

[2] R. Bouc, and all.
Hilbertian unilateral problems in viscoelasticity
Lecture Notes in Mathematics, n° 503, pp.219-234, 1975.

[3] R. Bouc, and all.
Cauchy and Periodic Unilateral Problems for Aging Linear Viscoelastic Materials
Jour. of Math. Analysis and Applications vol. 61, n°1, November 1977.

Dynamic Analysis of Structural Problems with Strong Physical Non-Linearities Described by Superpotentials

P.D. PANAGIOTOPOULOS

School of Technology
Aristotle University, Thessaloniki,
Greece.

Summary

The purpose of the present research is the derivation of variational inequalities, describing the dynamic behaviour of structures presenting strong physical nonlinearities and discontinuities as jumps, locking effects, etc. To treat the arising nonlinearities the notion of "superpotential" is used. The respective identification problem is also studied and some propositions are given.

The phenomenon of finite discontinous deformations called, Masson-Savart effect [1] has been studied experimentally by many research workers. In the metal alloys the phenomenon is due to possible material impurities, in other materials as polymers, rock masses, etc. it is caused by locking effects, cracks etc. Another kind of discontinuity effects arises when friction or non-bonding contact conditions hold on the boundary of the body. Assuming that the discontinuities are of monotone nature ([1] Fig.4.18.1-2) we desribe them by means of multivalued law of the form

$$\boldsymbol{\sigma} \in \partial w(\boldsymbol{\varepsilon}) \quad \text{resp.} \quad \dot{\boldsymbol{\sigma}} \in \partial w(\dot{\boldsymbol{\varepsilon}}) \tag{1}$$

for a holonomic (resp. on incremental) problem where $\boldsymbol{\sigma}$, $\boldsymbol{\varepsilon}$ (resp. $\dot{\boldsymbol{\sigma}}$, $\dot{\boldsymbol{\varepsilon}}$) are the stress and strain tensor (resp. their increments) $w: \mathbb{R}^6 \to (-\infty, +\infty]$ is a convex, lowersemicontinuous and proper functional [2], the so-called superpotential [3], and ∂ denotes the subdifferentiation. Similarly we assume that between the boundary tractions $\boldsymbol{S}$ and the boundary displacements $\boldsymbol{u}$ (resp. their increments) a relation of the form

$$-\boldsymbol{S} \in \partial j(\boldsymbol{u}) \quad \text{resp.} \quad -\dot{\boldsymbol{S}} \in \partial j(\dot{\boldsymbol{u}}) \tag{2}$$

holds. The structure is discretized by an appropriate finite element scheme. It can be shown [4] that the dynamic behaviour of the structure is described by the variational inequality

$$W(\boldsymbol{v})-W(\boldsymbol{u}(t))+\phi(\boldsymbol{v})-\phi(\boldsymbol{u}(t))\geq\left(\boldsymbol{p}-\boldsymbol{M}\frac{d^2\boldsymbol{u}(t)}{dt^2}\right)^T(\boldsymbol{v}-\boldsymbol{u}(t))\ \forall\boldsymbol{v}\in R^n \quad (3)$$

where $\boldsymbol{p}$ and $\boldsymbol{u}$ are the load and displacement vectors $\boldsymbol{M}$ is the mass matrix and $W(\boldsymbol{u})=\{\sum_{\alpha=1}^{\nu} w_\alpha(\boldsymbol{e}_\alpha-\boldsymbol{e}^o_\alpha)$ if $\boldsymbol{e}=\boldsymbol{G}^T\boldsymbol{u}$ and $w_\alpha(\boldsymbol{e}_\alpha-\boldsymbol{e}^o_\alpha)<\infty\ \forall\alpha$, ∞ otherwise$\}$ and $\phi(\boldsymbol{u})=\{\sum_{\beta=1}^{\nu'} f_\beta(\boldsymbol{u}_\beta)$ if $f_\beta(\boldsymbol{u}_\beta)<\infty$, ∞ otherwise$\}$

In the foregoing relation α (resp. β) denotes a node of the body (resp. of its boundary part where (2) holds), $\boldsymbol{G}^T$ is the compatibility matrix and $\boldsymbol{e}^o$ is the initial strain vector. By definition (3) can be written in the form of a differential inclusion

$$\boldsymbol{p}-\boldsymbol{M}\frac{d^2\boldsymbol{u}(t)}{dt^2}\in\partial W(\boldsymbol{u}(t))+\partial\phi(\boldsymbol{u}(t)) \quad (4)$$

Moreover the initial conditions $\boldsymbol{u}(t=0)=\boldsymbol{u}_0$, $\frac{d\boldsymbol{u}}{dt}(t=0)=\boldsymbol{u}_1$ hold.

Relation (4) holds also for the case of non-convex functionals. After Clarke [5] we define the generalized gradient as

$$\bar{\partial}W(\boldsymbol{u})=\{\boldsymbol{s}/(\boldsymbol{s},-1)\in K(\boldsymbol{u},W(\boldsymbol{u}))\}$$

where K denotes the normal cone to the set "epigraph W " at the point $\boldsymbol{u}$, and W again a lower semicontinuous and proper functional. Similarly for ϕ . Then (4) holds with ∂ replaced by $\bar{\partial}$. In order to solve (3) numerically a time disretization scheme is used and in every time increment Δt a static variational inequality equivalent to a minimization problem of the general form

$$\frac{1}{2}\boldsymbol{U}^T\tilde{\boldsymbol{M}}\boldsymbol{U}+W(\boldsymbol{U})+\phi(\boldsymbol{U})-\tilde{\boldsymbol{P}}^T\boldsymbol{U}\rightarrow\min \quad (5)$$

will be solved. Here $\boldsymbol{U}$ denotes the displacements at the end of the considered time step and $\tilde{\boldsymbol{M}}$ (resp. $\tilde{\boldsymbol{P}}$) is a positive definite matrix (resp. a vector) depending on the time integration scheme.

Due to the discontinuity effects, the parameter identification [6] is of importance for such a structure. We introduce a control vector $\mathbf{z} \in U_{ad} \subset R^m$, where U_{ad} is convex and closed and the observation vector $h \in R^p$. The performance index is

$$R(\mathbf{z}) = \int_0^T \| M u(\mathbf{z},t) - h \|^2 dt \qquad (6)$$

where $M: R^n \to R^p$ is a given matrix. To every control $\mathbf{z}$ a state function $u(\mathbf{z})$ is associated by means of the variational inequality (3) with p replaced by $p+B\mathbf{z}$, and $u(t)$ by $u(\mathbf{z},t)$, where $B: R^m \to R^n$. The identification problem has the form: Find $J \in U_{ad}$ such that

$$R(J) = \inf \{ R(\mathbf{z}) / \mathbf{z} \in U_{ad} \} \qquad (7)$$

The question of the existence of the solution is still open. We can propose two methods for the numerical treatment: The first consists in the consideration of the respective regularized problem. It results by replacing the generally non-differentiable functionals W and ϕ by a sequence of differentiables ones W_ρ and ϕ_ρ depending on a parameter $\rho > 0$ such that $W_\rho(v) \to W(v) \; \forall v \in R^n$ as $\rho \to 0$.

The regularized variational inequality (put in (3) W_ρ, ϕ_ρ, $p+B\mathbf{z}, u_\rho(\mathbf{z},t)$), is equivalent to the equation

$$\frac{dW_\rho(u_\rho(\mathbf{z}))}{du_\rho} + \frac{d\phi_\rho(u_\rho(\mathbf{z}))}{du_\rho} = p + B\mathbf{z} - M\frac{d^2 u_\rho(\mathbf{z},t)}{dt^2} \qquad (8)$$

and thus we have to solve a classical identification problem. Its solution $u_\rho(J_\rho,t), J_\rho(t)$ for $\rho \to 0$ approximates (heuristic method) the solution of the initial problem. The other method consists in the time discretization and the solution of the identification problem in every time increment. Within each step a variational inequality of the form

$$U^T\tilde{M}(V-U)+W(V)-W(U)+\phi(V)-\phi(U)-$$
$$-(\tilde{P}+BZ)(V-U)\geq 0 \quad \forall V\in R^n \qquad (9)$$

relates the state to the control. Arguing as in [6] (prop.1,5) we can prove that on the assumption a) $int\, D_W \cap int\, D_\phi \neq \emptyset$ and b) U_{ad} bounded, the time increment- identification problem has at least one solution. Moreover assuming for W_ρ (and for ϕ_ρ) that a) liminf $W_\rho(U_\rho)\geq W(U)$ for $U_\rho \to U$ and $\rho \to 0$ and b) that there exists a bounded sequence $\{\bar{U}_\rho\}$ such that $(dW_\rho(\bar{U}_\rho)/dU_\rho)^T U \leq C\|U\|$ $C>0$ constant, the solution Z , U , R of the regularized identification problem within each increment tends to the solution Z , U , R of the initial problem.

References

[1] Bell,J.F.: The Experimental Foundations of Solid Mechanics. Encyclopedia of Physics Vol.VIa/1 (ed.C.Truesdell). Berlin, Heidelberg: Springer-Verlag (1973).

[2] Rockafellar,R.T.: Convex Analysis. Princeton Univ.Press. Princeton, New Jersey (1970).

[3] Moreau,J.J.: La notion de sur-potentiel et les liaisons unilatérales en élastostatique.C.R.Acad.Sc.267A.Paris (1968) 954-957.

[4] Panagiotopoulos,P.D.: Optimal Control of Unilateral Structural Analysis Problems. Structural Control (ed.H.H.E.Leipholz). North Holland Pub.Co. and SM Publ. (1980).

[5] Clarke,F.H.: Generalized gradients and applications. Trans.Amer. Math.Soc.205 (1975) 247-262.

[6] Maier,G.: Future Directions in Engineering Plasticity. Engineering Plasticity by Mathematical Programming (eds. M.Z.Cohn, G. Maier). New York, Toronto, Paris: Pergamon Press (1979).

A Quadratic Approximation of a Non-Linear Structural Design Problem

G. PAPE and G. THIERAUF

Essen University
4300 Essen, Fed. Rep. of Germany

1. Introduction

Optimality of structural systems has been a main field of interest in the past ten or twenty years. The present article summarizes one aspect of optimal design in the plastic range, which is firmly rooted in the well-known Prager-Shield-Theory of optimal design /1/. The derivation given here is limited in space; for a more detailed discussion we refer to /2/ and /4/. Applications of the Prager-Shield-Theory in analytical form can be found in /3/.

2. A Nonlinear Optimization Problem in Plasticity

For structures which can be analysed by means of a finite element discretization, we formulate the following Problem:

$$\begin{aligned} &\text{Minimize } Z(\underline{y}) , && \underline{y} = \{y_1, \ldots, y_r\} && (1) \\ &\text{subject to} && && \\ &\text{(yield condition)} && \psi_i(\underline{F}_i^j) \leqq y_i , && i = 1, \ldots, r \\ &\text{(equilibrium)} && \underline{N}\,\underline{F}^j = \underline{P}^j , && j = 1, \ldots, k \end{aligned}$$

$\underline{F}^j$ is the vector of linear independent internal forces in equilibrium with loading case $\underline{P}^j$, $\underline{N}$ is the equilibrium matrix of the structure and ψ_i is a yield condition or constraint on the internal forces, in general nonlinear but convex and piece-wise differentiable.
The objective function $Z(\underline{y})$ relates to the yield condition in the design variables $\underline{y}$; this is a restriction to nonhardening plastic behaviour. With regard to a most simple structural system (fig. 1) equations (1) describe the following

design problem:
For two loading cases P_1 and P_2 the minimum of a design function has to be found. The constraints are formulated in the internal forces $\underline{F}$ by a nonlinear yield function defined in an appropriate number of points (i). There is no restriction in the number of control points i per element or in the number of design variables per element.

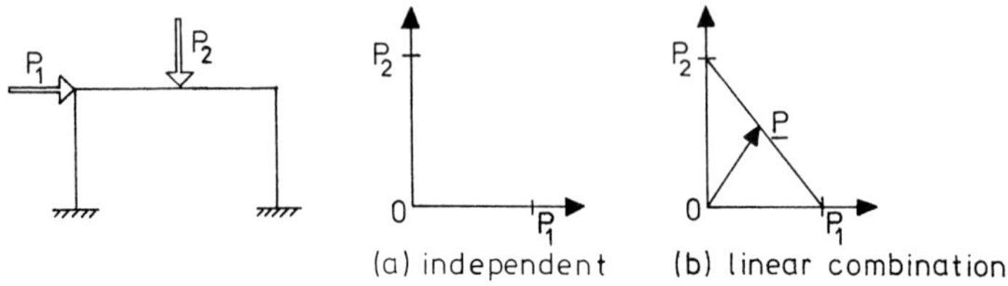

Fig. 1: Plane frame under two loading cases

Extensions are minimum values for y_i, symmetry conditions and deformation constraints formulated for an associated elastic structure. Convex combinations of the loading cases as indicated in fig. 1 are included.

3. Iterative Solution by Quadratic Approximation

In contrast to comparable solutions by linearization, we have chosen a quadratic approximation which relates in a natural way to the basic finite element solution technique. It is described in short by a simplified, but basic problem which constitutes the general iteration as described in /2/:

$$\text{Minimize} \quad Z^{*}(\underline{y}) = c\ y \ \text{(one variable)} \tag{2}$$

$$\text{subject to} \quad \underline{F}^T \underline{A}_1 \underline{F} \leq y\ ,$$

$$\underline{F}^T \underline{A}_2 \underline{F} \leq y$$

$$\text{and} \quad \underline{N}\ \underline{F} = \underline{P} \ \text{(one loading case)},$$

where $\underline{A}_i$ is symmetric and positive definite.

We formulate the Lagrangian L and the corresponding Kuhn-Tucker conditions (comma for part. derivatives):

$$L = c\ y + \underline{u}^T (\underline{N}\ \underline{F} - \underline{P}) + \sum_i \lambda_i \left(\underline{F}^T \underline{A}_i \underline{F} + \alpha_i^2 - y\right) \tag{3}$$

and $$L,_y \equiv c - \sum_i \lambda_i = 0 \tag{4}$$

$$\left.\begin{aligned} L,_{\underline{u}} &\equiv \underline{N}\,\underline{F} - \underline{P} = \underline{0} \\ L,_{\underline{F}} &\equiv \underline{N}^T\,\underline{u} + 2\sum_i \lambda_i\,\underline{A}_i = \underline{0} \\ L,_{\lambda} &\equiv \underline{F}^T\,\underline{A}_i\,\underline{F} + \alpha_i^2 - y = 0 \\ L,_{\alpha} &\equiv \lambda_i\,\alpha_i = 0 \end{aligned}\right\} \text{for } i = 1, 2 \tag{5}$$

λ_i and u_j are Lagrange parameters, α_i are slack-variables. From (4) and (6) we see that

$$0 < \lambda_i < c$$

and from the principle of minimum potential energy we can conclude (see /2/) that there must be an associated elastic structure with forces $\underline{F}$ that satisfy the strict yield condition in (5), namely

$$\underline{F}^T\,\underline{A}_i\,\underline{F} = y \tag{6}$$

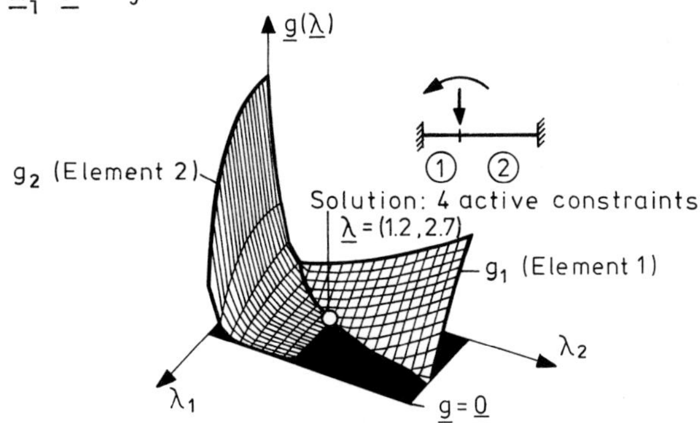

Fig. 2: Seperation into active and passive constraints

In the present case this results in a nonlinear condition g for λ_i, which is plotted in fig. 2. The cutting points with the λ_1,λ_2-plane locate the Lagrange parameters which is identical to a seperation into active and passive constraints.

For the original problem, this iteration is performed in an inner loop (fig. 3), an outer loop serves for a linearization of Z and for the basic quadratic approximation of ψ, resulting in the matrices $\underline{A}_i$ as used in equations (5) and (6). Typical iteration histories are shown in fig. 3.

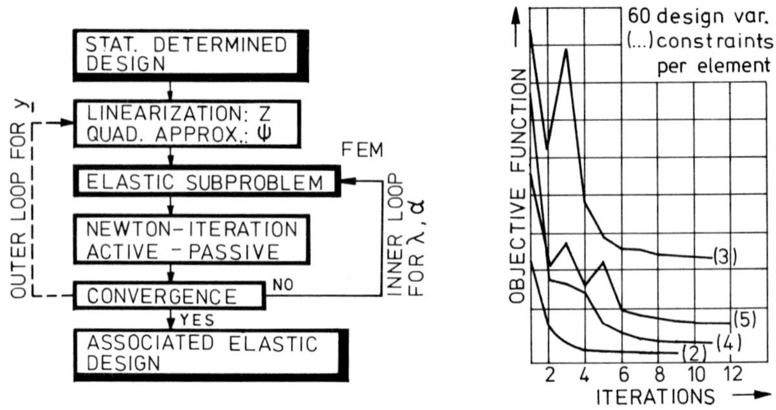

Fig. 3: FLOWCHART and typical iteration histories

Two sample problems are given in fig. 4, including some specific results for minimum weight design and for several loading cases.

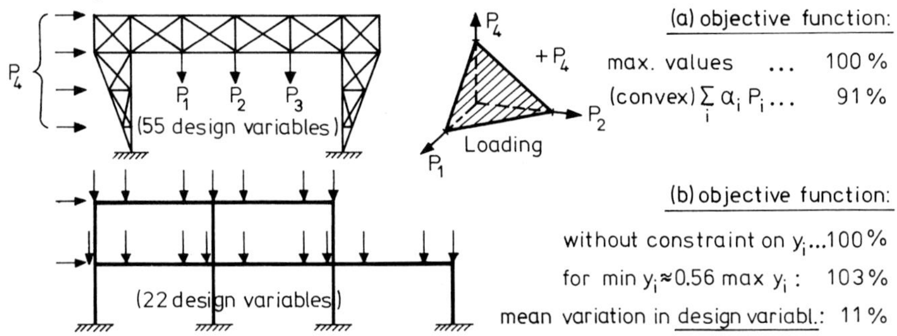

Fig. 4: Examples for convex load combinations (a) and minimum design variables (b)

References

/1/ PRAGER, W. and SHIELD, R.T., "A General Theory of Optimal Plastic Design", J. of Appl. Mech., 1967, pp. 184-186.

/2/ PAPE, G., "Eine quadratische Approximation des Bemessungsproblems idealplastischer Tragwerke", Ph. D. Dissertation, Essen University, 1979.

/3/ ROZVANY, G.I.N., "Optimal Design of Flexural Systems", Pergamon Press, Oxford, 1976.

/4/ THIERAUF, G., "A Method for Optimal Limit Design of Structures with Alternative Loads", Comp. Meth. in Appl. Mech. and Eng., Vol. 16, 1978, pp. 135-149.

Calculation of Viscoelastic Response

Y. Partom

A.D.A., P.O.B. 2250, Haifa, ISRAEL.

Summary

We show that a grains model (GRM) constitutive description is capable to account in part for the main features of viscoplastic response. This provides us with a physical basis for using the phase model (PHM) constitutive description, originally developed for nonlinear viscoelastic response, to describe viscoplastic response as well. We calibrated the PHM flow diagram to match stress-strain curves for titanium and then computed various response curves. Because of lack of space we show only the cyclic loading response curve.

Introduction

In several papers during the last ten years Bodner and Partom [1, 2, 3] developed a constitutive representation for viscoplastic response, sometimes referred to as the Bodner-Partom model (BPM). The BPM is suitable for numerical integration of boundary value problems and has been applied to different materials and to different loading configurations.

In terms of mechanical models, usually used to represent viscoelastic response, the BPM is a Maxwell model with a nonlinear dashpot as shown in Fig. 1. A nonlinear dashpot is defined by a flow law $\dot{\varepsilon}^p(\sigma)$ where for a linear dashpot we have as usual $\dot{\varepsilon}^p = \sigma/\eta$. In the BPM the flow law contains the additional internal variable Wp (plastic work) so that we have a continuous family of flow curves for the same dashpot. The additional

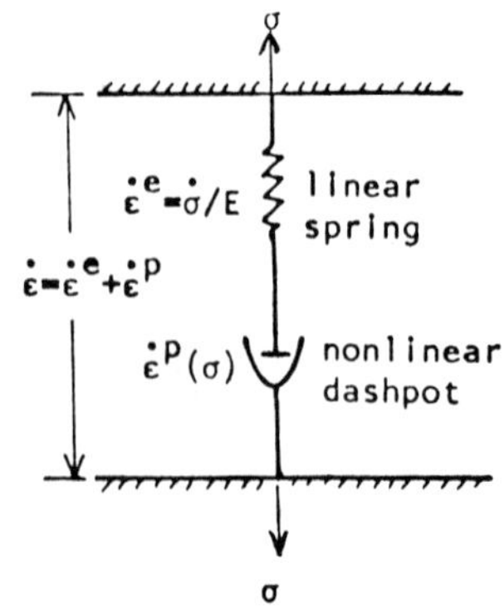

Fig. 1 - A Maxwell Model with a Nonlinear Dashpot.

variable was introduced mainly to account for the strain hardening effect. Being a single element Maxwell model the BPM is, however, limited in its ability to reporduce a wide spectrum of realistic material response. From [3] we see that it was not quite successful when applied to reverse and cyctic loadings with a Bauschinger effect.

Recently we used a generalized Maxwell model with nonlinear dahspots, which we call the phases model (PHM), to represent realistic nonlinear viscoelastic response [4]. While doing this we realized that such a model reproduces in a natural way modes of behavior similar to strain hardening and Bauschinger's effect. An equivalent approach to viscoplastic response was used by Iwan [5] and by Wells and Paslay [6].

It is usually accepted that the main features of viscoplastic response, i.e., strain hardening and Bauschinger's effect result from variations in the dislocations configuration [7]. There is however a line of authors who proposed that these types of response can also result from the grain structure of a polycrystalline material. We refer to this approach as a grains model (GRM). The first to propose a GRM seems to be Duwez [8] before dislocations were invented. He was followed much later by a line of authors such as Budiansky and Wu [9], Hill [10], and Brown [11], and more recently by Lin, Yu and Weng [12], Weng [13] and Patel and Bieniek [14]. The GRM concerns us here mainly because it provides a physical basis for applying the PHM to viscoplastic response. Each phase of the PHM can be thought of as representing the response of several grains lumped together.

In what follows we apply the GRM, in one of its simplest forms, to the constant rate unaxial stress loading of titanium, and calibrate our PHM to match the same curves. We applied the PHM to various loading paths, but because of lack of space we show only the calculated response to cyclic loading.

The grains model (GRM)

We assume that a volume element of a polycrystalline material contains many identical single crystal grains randomly oriented in space. Each such grain has a slip system (or several identical slip systems) for dislocations motion. In its slip plane the crystal behaves like a Maxwell model with a nonlinear dashpot similar to the BPM. The slip plane flow law is defined by a $\dot{\gamma}^p(\tau)$ relation where $\dot{\gamma}^p$ is the plastic resolved strain rate and τ is the resolved shear stress for the slip system. For the elastic part of the response we assume, as in [14], that the grains are isotropically elastic with the constants of the polycrystal. To account for the combined response of the aggregate we assume that all the grains have the same strain equal to the macroscopic strain. This non self-consistent approach, which is easy to handle, was shown by Patel and Bieniek [14] to be a good approximation to a fully self-consistent approach. The GRM equations according to these assumptions and for the special case of unaxial stress are as follows:

Let x be the unaxial stress direction and (p,q) the slip system axes for one of the grains. The transformation from (x,z) to (p,q) is achieved by two rotations. First rotate (x,z) by ϕ around y to get (p,z'). Then rotate z' by ψ around p to get (p,q). It can be shown that $\dot{\gamma}^p$ and τ are given by:

$$\tau=\sigma_{pq}=\tfrac{1}{2}\cos 2\phi\cos\psi\sigma_{xx}=f(\phi,\psi)\sigma_{xx} \tag{1}$$

$$\dot{\gamma}^p=\dot{\varepsilon}^p_{pq}=\frac{2}{3}f(\phi,\psi)\dot{\varepsilon}^p_{xx} \tag{2}$$

To compute a constant rate $\dot{\varepsilon}_{xx}=\dot{\varepsilon}$ response we increase the strain in small steps $\Delta\varepsilon$. We divide the full range of the rotation angles into n equal steps so that we have n^2 different orientations, and assume that the slip system of one of the grains coincides with each such orientation. In the actual computations we used $n=20$ for a $\pi/2$ range. For each grain we store the value of τ at the previous step (denoted by τ^k). To compute σ^{k+1}_{xx} we solve a nonlinear equation for each grain as follows:

$$\dot{\varepsilon}^e_{xx}=\left(\frac{\tau^{k+1}-\tau^k}{E\Delta t}\right)/f=G(\tau^{k+1}) \tag{3}$$

$$\dot{\varepsilon}^p_{xx}=\dot{\gamma}^p(\tau^{k+\frac{1}{2}})/\tfrac{2}{3}\,f=H(\tau^{k+1}) \tag{4}$$

$$G(\tau^{k+1})+H(\tau^{k+1})=\dot{\varepsilon} \tag{5}$$

$$\sigma^{k+1}_{xx}=\frac{1}{n^2}\sum_{\substack{\text{all}\\ \text{grains}}}\tau^{k+1}/f \tag{6}$$

In Fig. 2 we show the results obtained from the GRM when applied to the experimental curves for titanium taken from [2]. The flow law we used was:

$$\dot{\gamma}^p=\dot{\gamma}_0(e^{\beta\tau}-1)$$
$$\dot{\gamma}_0=1\times10^{-18}\text{sec}^{-1};\quad \beta=53.2\text{Kbar}^{-1} \tag{7}$$

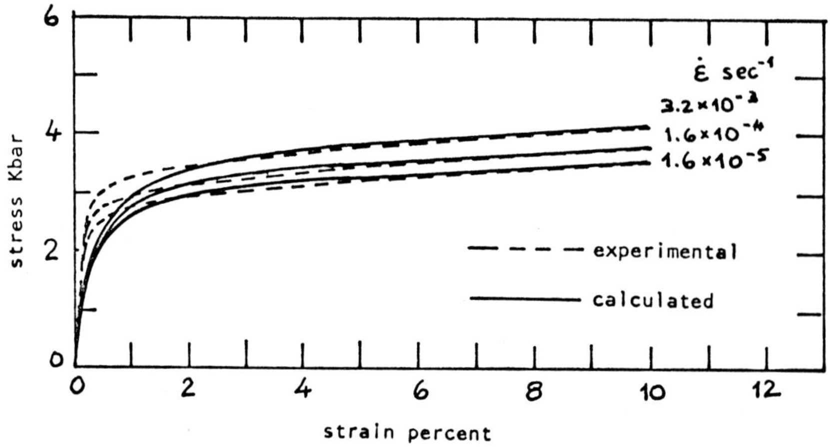

Fig. 2 - Constant Rate Curves for Titanium. Experimental and Calculated by the GRM.

which we arrived at in two steps: We first used an elastic-perfect-plastic type flow law to determine a yield stress τ_y for each curve separately. We than obtained a first approximation to Eq. (7) from a semi-logarithmic plot of the pairs $(\dot{\varepsilon}, \tau_y)$.

From Fig. 2 we see that the agreement with the experimental curves is generally quite good, but there is some disagreement at the knee of each of the curves. Notice in particular that a strain hardening behavior was obtained without recourse to an additional internal state variable. We are not using the GRM in other more complicated types of loading for two reasons. First, because the GRM computations are very demanding in computer time and storage and would become unpractical even for 1-D time dependent boundary value problems. This is because a minimum of 200 grains is needed to get reasonable representation of grains orientations. Second, because the GRM is still in a stage where it requires some refinement to be able to produce realistic results. We therefore regard the GRM mainly as an approach that provides us with a physical basis for applying the PHM to metals.

The phases model (PHM)

The PHM can be looked upon as a generalized Maxwell model with nonlinear dashpots. Each Maxwell element is refrred to as a *phase*. The numbers of phases needed to get realistic results is of the order of ten. Each phase can be thought of as representing the combined response of several grains lumped together. Each phase has its own flow law and the plot of all these flow curves is called the flow diagram. We calibrate the flow diagram to fit a family of constant rate stress-strain curves. The procedure of calibration is quite straightforward, but we are not able to describe it here because of lack of space. For the example given below we used ten phases and calibrated the flow diagram against three experimental constant rate curves for titanium form [2]. The reproduced curves are shown in Fig. 3 together with the experimental curves and we see that the agreement is quite good. With the same

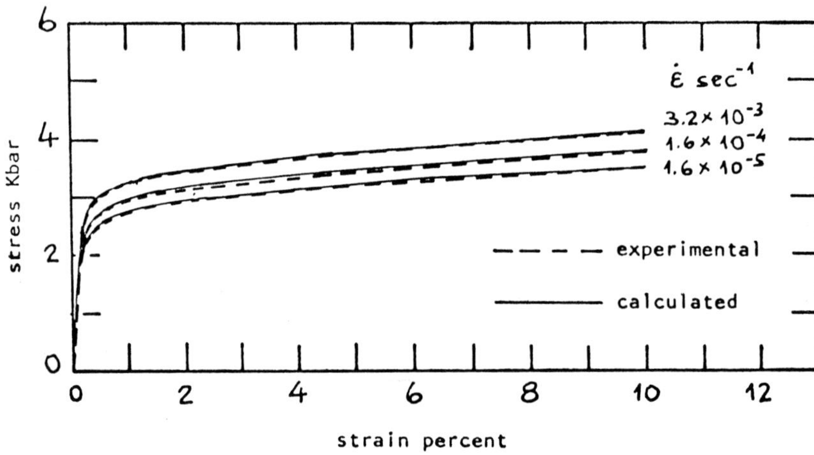

Fig. 3 - Constant Rate Curves for Titanium. Experimental and Calculated by the PHM.

flow diagram we calculated the response to other loading paths and obtained reasonable results. Because of lack of space we show in Fig. 4 only the cyclic loading curve. An experimental cyclic loading curve for titanium is given in [3].

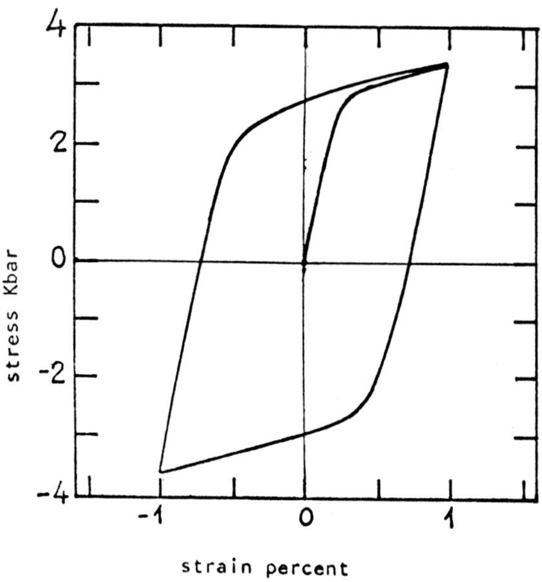

Fig. 4 - Cyclic Loading to ±1% for Titanium Calculated by the PHM.

Comparing the two we conclude that the agreement between the calculated and experimental results is quite reasonable. They differ, however, in that subsequent calculated cycles coincide while there is a small drift between subsequent experimental cycles. This drift has to be a result of variations in the dislocations configuration.

Conclusions

The important features of viscoplastic response of polycrystalline metals are due in part to the combined effect of single crystal grains randomly oriented in space. The grains model (GRM) which we presented is an approximate constitutive representation of this effect. Careful tests on the response of single crystals versus polycrystals are needed to determine exactly which part of the viscoplastic response is due to the grains effect and which part is caused by changes in the internal dislocated state.

References

1. Bodner, S.R.; Partom, Y.: A large deformation elastic-viscoplastic analysis of a thick walled spherical shell. J. Appl. Mech., 39 (1972), 571-757.

2. Bodner, S.R.; Partom, Y.: Constitutive equations for elastic-viscoplastic strain hardening materials. J. Appl. Mech. 42 (1975) 385-389.

3. Bodner, S.R.; Partom, I; Partom, Y.: Unaxial cyclic loading of elastic-viscoplastic materials. J. Appl. Mech. 46 (1979) 805-810.

4. Partom, Y.; Schanin, J.: Calculation of nonlinear visco-elastic response. M.Sc. thesis, to be published.

5. Iwan, W.D. : On a class of models for the yielding behavior of continuous and composite systems. J. Appl. Mech. 34 (1967) 612-617.

6. Wells, C.H.; Paslay, P.R.: A small strain plasticity theory for planar slip materials. J. Appl. Mech. 36 (1969) 15-21.

7. Cottrell, A.H.: Dislocations and plastic flow in crystals. Oxford at the Clarendon Press (1965).

8. Duwez, P. : On plasticity of crystals. Phys. Rev. 47 (1935) 494-501 .

9. Budiansky, B.; Wu, T.T.: Theoretical prediction of plastic strains in polycrystals. Proc. 4th U.S. Nat. Cong. Appl. Mech. (1962) 1175-1185.

10. Hill, R. : Continuum micro-mechanics of elastoplastic polycrystals. J. Mech. Phys. of Solids, 13 (1965) 89-101.

11. Brown, G.M.: A self-consistent polycrystalline model for creep under combined stress states. J. Mech. Phys. of Solids 18 (1970) 367-381.

12. Lin, T.H.; Yu, C.L.; Weng, G.J.: Derivation of polycrystal creep properties from the creep data of single crystals. J. Appl. Mech. 44 (1977) 73-78.

13. Weng, G.J.: A physically consistent method for the prediction of creep behavior of metals. J. Appl. Mech. 46 (1979) 800-804.

14. Patel, M.R.; Bieniek, M.P.: An analysis of the viscoplastic behavior of metals. Mat. Sci. and Engrg. 40 (1979) 123-134.

Calcul par Elements Finis de l'Evolution d'une Inclusion dans une Matrice Deformée Uniformement

J.PERROT, J.L. CHENOT

CEMEF - ENSMP - ERA 837
SOPHIA ANTIPOLIS - VALBONNE

Summary

A new finite element numerical model is presented, for predicting the occurence of micro voids near inclusions. Materials rheology is Norton-Hoff law, mechanical description is Eulerian with incompressibility and Newton method is used together with ε-algorithm for calculating the numerical solution.

Introduction

Malgré les efforts qui y sont consacrés, les phénomènes d'endommagement dans les métaux ne sont pas encore bien élucidés. De nombreux modèles ont été mis au point, pour étudier la croissance et la coalescence de trous microscopiques : analytique [1] ou par éléments finis [2]. Le problème de l'amorçage des micro-fissures a été surtout étudié pour des matériaux élastiques, ou élasto-plastique [3] et pour de petites déformations. Aussi pour les problèmes de ductilité en mise en forme à chaud, des métaux présentant des inclusions, où les déformations sont importantes un nouveau modèle numérique s'avère nécessaire pour prévoir l'apparition de défauts.

1. Formulation du problème

On se limite au problème à deux dimensions : l'inclusion est au centre d'un carré figurant les limites de la matrice, et a une section initialement circulaire (figure 1).

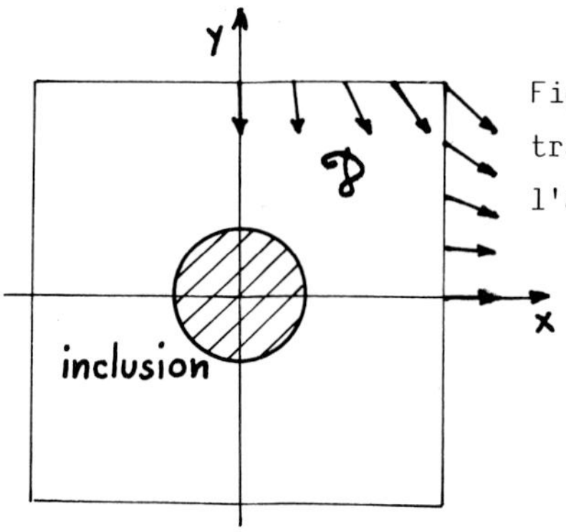

Figure 1:Schématisation du problème: traction uniforme de la matrice selon l'axe Ox.

Le comportement de la matrice et de l'inclusion est décrit par une loi de Norton-Hoff [4] du type :

$$[s] = 2K \frac{[\dot{\varepsilon}]}{(\sqrt{3}\ \dot{\bar{\varepsilon}})^{1-m}} \qquad \text{où } \dot{\bar{\varepsilon}} = \left(\frac{2}{3} \sum_{i,j} \dot{\varepsilon}_{ij}^{2}\right)^{1/2} \qquad (1)$$

([s] est le déviateur des contraintes, [$\dot{\varepsilon}$] le tenseur des vitesses de déformation, K et m sont des constantes éventuellement différentes pour la matrice et l'inclusion).

On peut introduire un écrouissage isotrope par

$$K = K\,(\bar{\varepsilon}) \qquad \text{où } \bar{\varepsilon} = \int_{\widehat{MoM}} \dot{\bar{\varepsilon}}\,dt \qquad (2)$$

$\widehat{MoM}$ étant l'arc de courbe décrit par le point initialement en Mo pour la configuration initiale.

Compte tenu des conditions aux limites, portant uniquement sur les vitesses, soit $\vec{v}^d$, le champ de vitesse solution du problème vérifie

$$\operatorname{div} \vec{v} = 0$$
$$\vec{v} = \vec{v}^d \text{ sur les frontières}$$

et minimise la fonctionnelle

$$I(\vec{v}) = \frac{1}{m+1} \iint_{\mathcal{D}} K\,(x,y)\,(\sqrt{3}\,\dot{\bar{\varepsilon}}\,)^{1+m}\ dxdy \qquad (3)$$

(m est supposé ici identique pour la matrice et l'inclusion, K est différent et évolue en fonction de l'écrouissage).

Ayant choisi, à ce stade, une formulation purement eulérienne, les coordonnées $\vec{X}$, d'un point matériel, évoluent selon la loi :

$$\frac{d\vec{X}}{dt} = \vec{v} \qquad (4)$$

2. Dicrétisation en éléments finis

Si l'on utilise des éléments standards, on ne peut imposer qu'une incompressibilité moyenne dans l'élément, aussi avons nous préféré l'élément quadrangulaire subdivisé en 4 triangles, qui permet d'imposer l'incompressibilité exacte [5]. Nous avons préféré, à la formulation classique, utilisant les multiplicateurs de Lagrange identifiées aux pressions p [6], la pénalisation, plus économique, à partir de la fonctionnelle :

$$J(\vec{v}) = \frac{1}{m+1} \iint_{\mathcal{D}} K(x,y)(\sqrt{3}\,\dot{\bar{\varepsilon}})^{1+m} dxdy + \iint_{\mathcal{D}} \rho(\operatorname{div} \vec{v})^2 dxdy \qquad (5)$$

avec $\rho > 0$

Dans le problème discret nous avons un système d'équations différentielles à résoudre issu de (4), le champ de vitesse étant donné par la minimisation de (5). Le maillage est ainsi réactualisé à chaque étape de la déformation.

La pression hydrostatique est calculée à partir de la formule :

$$\overrightarrow{\text{grad}}\ p = \operatorname{div} [s]$$

3. Méthode de résolution

Nous avons testé la méthode de Newton sous incrémentée [7], pour minimiser (5), mais la convergence numérique, satisfaisante pour un matériau homogène avec $m \geq 0,1$, n'était plus assurée pour un matériau hétérogène. Nous avons donc dû faire appel aux techniques d'accélération de la convergence [8] ; l' ε-algorithme vectoriel donne des résultats très satisfaisants dans notre cas.

L'intégration par rapport au temps n'a pas encore été optimisée ; nous avons choisi dans un premier temps le schéma simple d'Euler, pour les déplacements et les déformations généralisées $\dot{\bar{\varepsilon}}$.

4. Résultats

Le modèle que nous avons décrit permet de calculer les contraintes, les déformations, et même les dommages cumulés, au sens de Latham et Cockcroft [9], pour une déformation quelconque.

A titre d'exemple, nous donnons ci-dessous (figure 2 à 5) les résultats de nos calculs, pour une traction simple selon l'axe Ox. Les données numériques sont :

matrice $K = 1 \quad m = 0,1$

inclusion $K = 1,5 \quad m = 0,1$

la loi d'écrouissage est dans les deux cas : $K = K_o\,\bar{\varepsilon}$

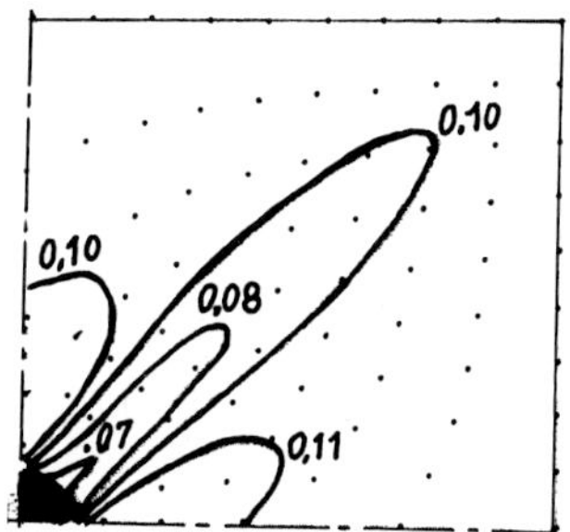

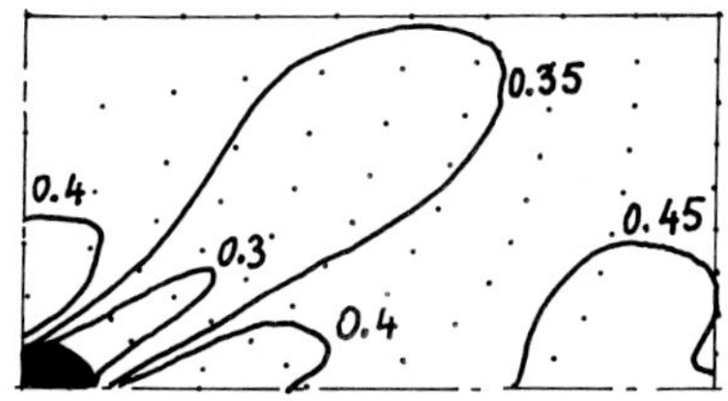

Figure 3 : Déformation 40 %
Figure 2 . Déformation 10 %

Cartes d'isodéformation généralisée

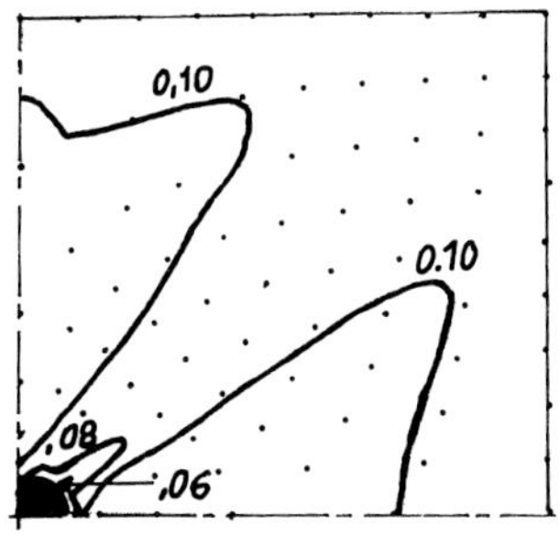

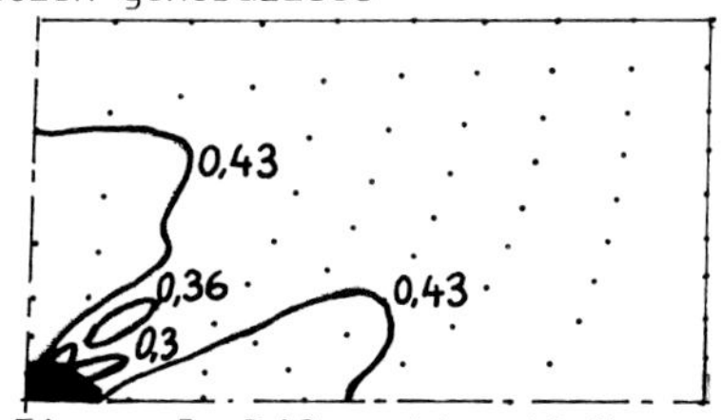

Figure 5: Déformation 10 %

Figure 4: Déformation 40 %

Cartes d'iso-endommagement (Latham et Cokcroft)

Conclusion

Le modèle présenté permet de prévoir l'apparition de décohésions, pour un chemin de déformation quelconque, et moyennant des hypothèses simples sur les conditions d'amorçage au niveau microscopique: déformation excessive dans la matrice, contrainte critique à l'interface, endommagement excessif.

Bibliographie

1. Mc CLINTOCK F.A., ARGON A.S. - Mechanical Behavior of Materials, p. 524, New York, Addison-Wesley 1966.
2. ANDERSSON H. Analysis of a Model for Void Growth and Coalescence ahead of a Moving Crack Tip - J. Mech. Phys. Solids 25(1977) 217
3. ARGON A.S. - Formation of Cavities from non Deformable Second Phase Particules in Low Temperature Ductile Fracture - Trans.ASME (1976)
4. HOFF N.J. - Approximate Analysis of Structures in the Presence of Moderatly Large Creep Deformations - Quart. Appl. Math. 12 (1954) 49
5. NATEGAAL J.C., PARKS1D.M., RICE J.R., - On Numerically Accurate Finite Element Solutions in the Fully Plastic Range.-Comput. Math. Appl.Mech. 4 (1974) 153
6. ZIENKIEWICZ O.C., GODBOLE P.N., - Viscous Incompressible Flow with Special Reference to non Newtonian (Plastic) Fluids, Finite Element in Fluids - vol 1 p. 25 London Wiley and Sons 1975
7. ORTEGA J.M., RHEINBOLDT W.C., - Iterative Solution of non-Linear Equations in Several Variables New York, Acad. Press 1972.
8. BREZINSKI C., - Algorithmes d'accélération de la convergence Etude Numérique - Paris, Ed. Technip. 1978.
9. COCKCROFT M.G. LATHAM D., - A Simple Criterion of Fracture for Ductile Metals N.E.L. Report n 240 Scotland 1966.

Creep and Plastic Ratchetting in Cyclically Thermally Loaded Structures

A.R.S. PONTER* and M.M. MEGAHED†

*Department of Engineering, University of Leicester, England.

†Department of Mechanical Engineering, University of Cairo, Egypt.

Summary

The behaviour of a simple two bar structure subjected to cyclic thermal loading is investigated both experimentally and theoretically for transitional conditions between creep and plasticity, using copper as a model strain-hardening material. It is found that a simple recovery model provides a close prediction of the observed behaviour. The corresponding rapid cycle solution gives a reasonable upper bound to deformation rates even when the loads pass outside the shakedown limit.

Introduction

The paper discusses a combined theoretical and experimental study of the creep and plastic behaviour of a simple two-bar structure subjected to a constant primary load and a cyclic thermal loading. The general form of the structure and the temperature histories are shown in Fig. 1. The purpose of this work is to attempt to understand the most important material and structural considerations which should be taken into account in the prediction of deformation and rupture life and hence generate results of sufficient simplicity for use in design.

The range of thermal loads considered were restricted by the experimental requirement that the stresses remain tensile, as the model structure consisted of two coupled uniaxial machines [1]. The thermal loading has the effect, in these circumstances, of enhancing maximum stresses and hence causes creep acceleration.

The material used was high purity copper, and test specimens for the gathering of basic data and in the two bar structure were taken from the same billet. The creep properties of copper are well understood and, as a result, considerable progress could be made in understanding the observed behaviour, without introducing some of the

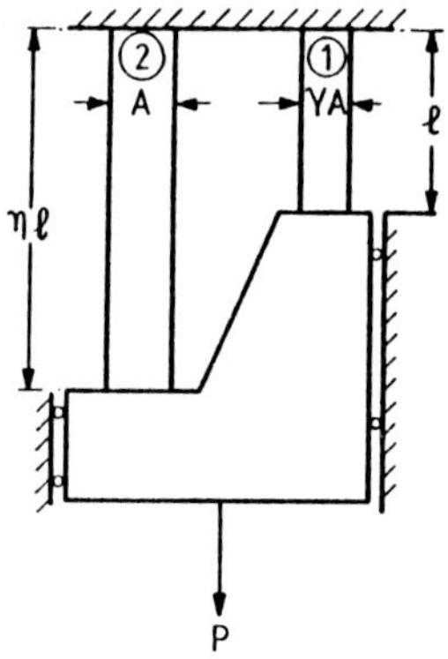

Fig. 1a Two bar structure

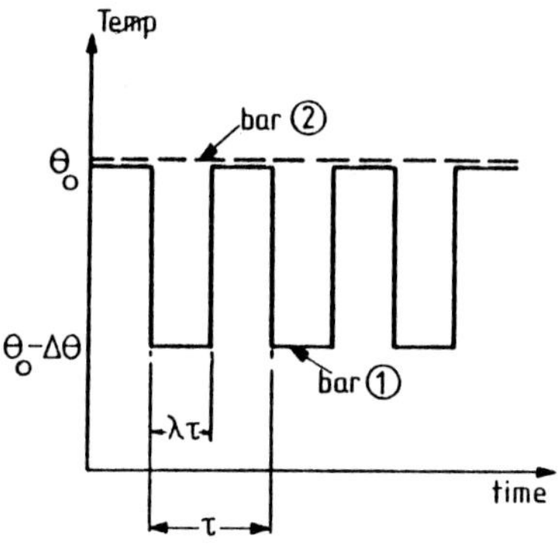

Fig. 1b Temperature history

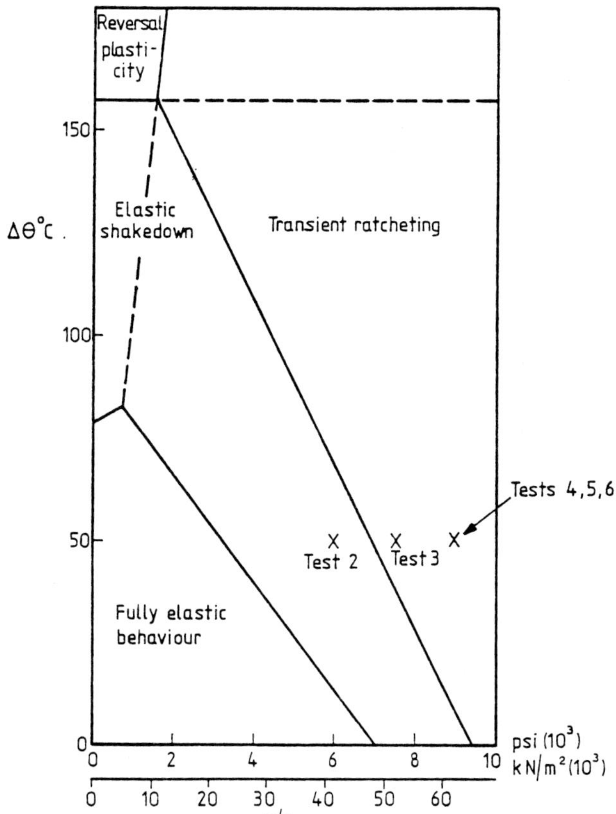

Fig. 2 Modes of deformation for a two bar structure with $\eta = 2$, $Y = 1$.
$\sigma y = 32 \cdot 10^3$ kN/m² (4650 psi)

Test	η	γ	θ_o °C	$\Delta\theta$ °C	P/A psi	τ hr	λ	Copper Batch
1	1	1	300	58*	8000	24	0.5	B2
2	2	1	300	50	5972	6	0.5	B2
3	2	1	300	50	7504	6	0.5	B2
4	2	1	300	58*→50	9042	6	0.5	B2
5	2	1	300	50	9005	24	0.5	B3
6	2	1	300	50	9037	48	0.5	B3

Table 1 Test conditions: (*an overshoot of 6°C and of duration ≈ 0.6 hr.)

additional difficulties associated with ferrous alloys. Using the extrapolated linear strain hardening curve to define an appropriate yield stress σ_y , which corresponded closely to the 0.1% proof stress, the Bree diagram shown in Fig. 2 can be constructed for the plastic behaviour of the structure.

The test details are given in Table 1 and it can be seen that the sequence of Tests 2 - 6 straddle the boundary between pure creep (Test 2) to a strong interaction between creep and plasticity (Tests 4 - 6). It is in this region that the material and structural behaviour is most difficult to understand and predict by simple means.

Material Models

The constant stress and temperature test data was modelled by the equations:

Plastic strains ε^p:

$$\varepsilon^p = 0 \qquad \sigma < \sigma_o \tag{1}$$

$$\varepsilon^p = \frac{B_o}{m+1}\left(\sigma^{m+1} - \sigma_o^{m+1}\right) \text{Exp}\left(\frac{-Q_h}{R\theta}\right) \qquad \sigma > \sigma_o \tag{2}$$

Steady-state creep stress rate $\dot{\varepsilon}^c$:

$$\dot{\varepsilon}^c = A_o \, \sigma^m \, \text{Exp}\left(\frac{-Q_c}{R\theta}\right) \tag{3}$$

where σ_o is the initial yield stress and B_o , m , Q_h , A_o n and Q_c are all material parameters found from plastic and creep tests covering the appropriate temperature and stress ranges. From this data two material models were constructed for variable stress and temperature. The simplest model (called viscous in the figures) ignored any possibility of interaction between the creep and plasticity and represents the total strain $\dot{\varepsilon}$ as

$$\dot{\varepsilon} = \frac{\dot{\sigma}}{E} + \varepsilon' = \frac{\dot{\sigma}}{E} + \dot{\varepsilon}^{p} + \dot{\varepsilon}^{c} \tag{4}$$

where $\dot{\varepsilon}^{p}$ is represented by an isotropically hardening model.

The second model involved interaction through recovery [2, 3, 4] where the inelastic strain rate $\dot{\varepsilon}'$ is given by

$$\dot{\varepsilon}' = f(\sigma - \bar{\sigma}) \tag{5}$$

where $\bar{\sigma}$ denotes an instantaneous flow stress;

$$f(\sigma - \bar{\sigma}) = 0 \;, \qquad , \; \sigma < \bar{\sigma} \;, \qquad\qquad > 0 \qquad , \; \sigma = \bar{\sigma} \;, \tag{6}$$

and $$\frac{d\bar{\sigma}}{dt} = \left[\frac{\partial\bar{\sigma}}{\partial\varepsilon}\right]_t \dot{\varepsilon}' + \left[\frac{\partial\bar{\sigma}}{\partial t}\right]_{\varepsilon'} \tag{7}$$

the first term in (7) representing strain hardening and the second recovery. The principal difference between the two models in that the recovery model predicts an instantaneous plastic strain when stresses rise to a repeated peak value. Such strains, which appear as a renewed transient primary creep strain, results from recovery of the dislocation structure and appears to form a feature of the deformation behaviour of most metals and alloys in the creep regime at stresses which approach the yield stress.

Theoretical Analysis

Complete step by step analyses were carried out for both material models, following as closely as possible the recorded temperature histories. In addition, rapid cycle solutions were constructed which, according to theory [3, 4] yield overestimate of the steady state deformation rates. In such solutions the stress history takes the form,

$$\sigma(t) = \hat{\sigma}(t) + \rho ,$$

where $\hat{\sigma}$ is the linearly elastic solution and ρ is a time constant residual stress field which is evaluated from compatability of accummulated inelastic strain over a cycle, assuming that the cycle time is small compared with characteristic material times.

Comparison of Theory and Experiment

Figure 3 shows both the experimental stress and deformation historic (σ_2 and ε_1) and the exact computed solutions for the two models for Test 3, and Fig. 4 summarizes the deformation rates results for three of the tests. In all cases the recovery model predicted quite closely the observed stress and strain histories whereas the simple creep/plasticity model underestimated the strain rate (due to the absence of recovery) and overestimated the amplitude of the thermal stresses (as the plastic strains due to recovery tend to reduce the stress applitude). As rupture approaches some stress redistribution takes place in the amplitude of stress variation although the maximum stress remains relatively constant. The stress amplitude depends upon the accummulation of strain per cycle and this increases as the material enters tertiary creep, causing the observed increase in minimum stress.

In Fig. 4 it can be seen that the rapid cycle solution for the recovery model (line c) is very close to the exact recovery solution (line a) and the test result for Test 2 which lies just within the shakedown limit, the circumstances which most closely mirrors a design situation. Divergence of the lines only become significant well outside the shakedown-limit. For test 4 failure occurred after 116 cycles with a failure strain in bar 1 of 9% corresponding to 1% strain in 13 cycles. For this test, use of the rapid cycle solution instead of the exact solution would imply a reduction in applied load P of about 14%. Hence the rapid cycle solution, using a recovery model can provide a safe estimate of deformation rates through a very simple solution method. Finite element methods for rapid cycle solutions are discussed by Ponter and Brown [25] and a complete description of the test results are

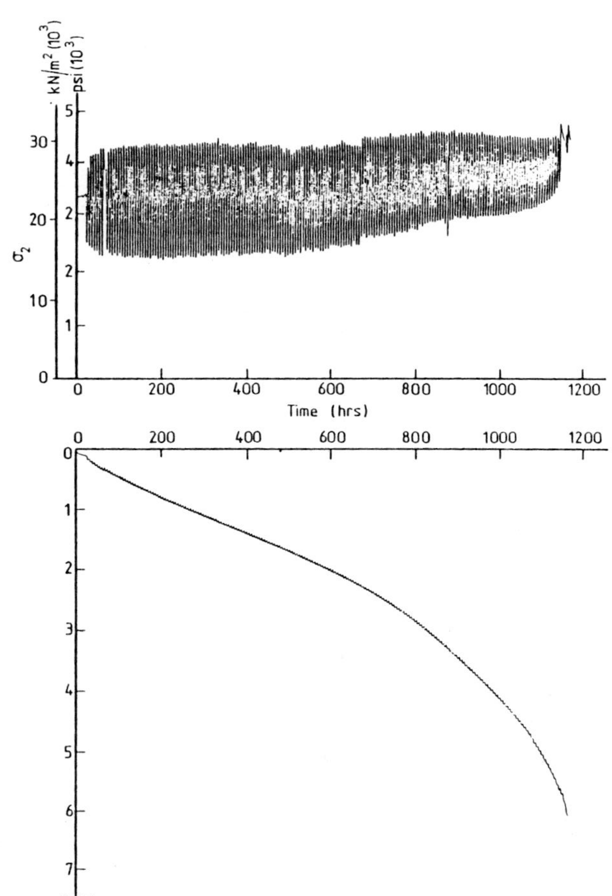

Fig. 3a Experimental results for test 3

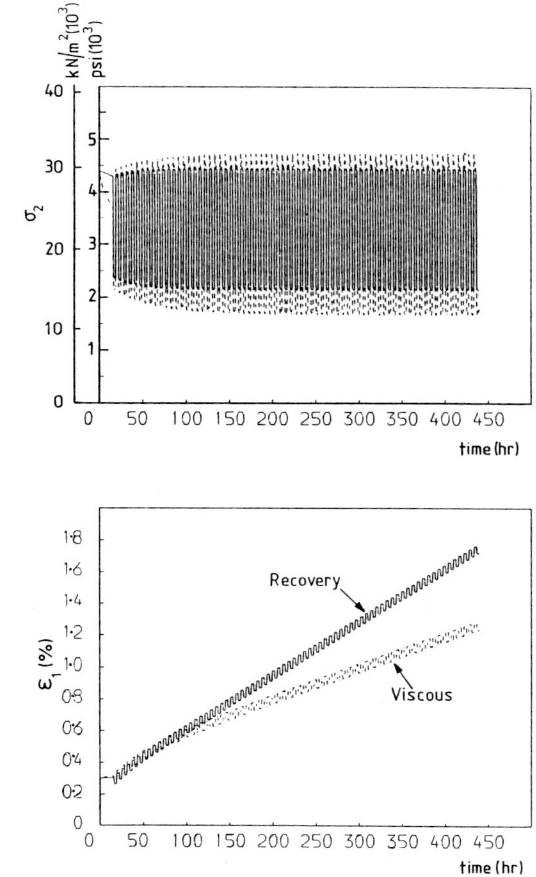

Fig. 3b Exact solutions for viscous and recovery models for Test 3

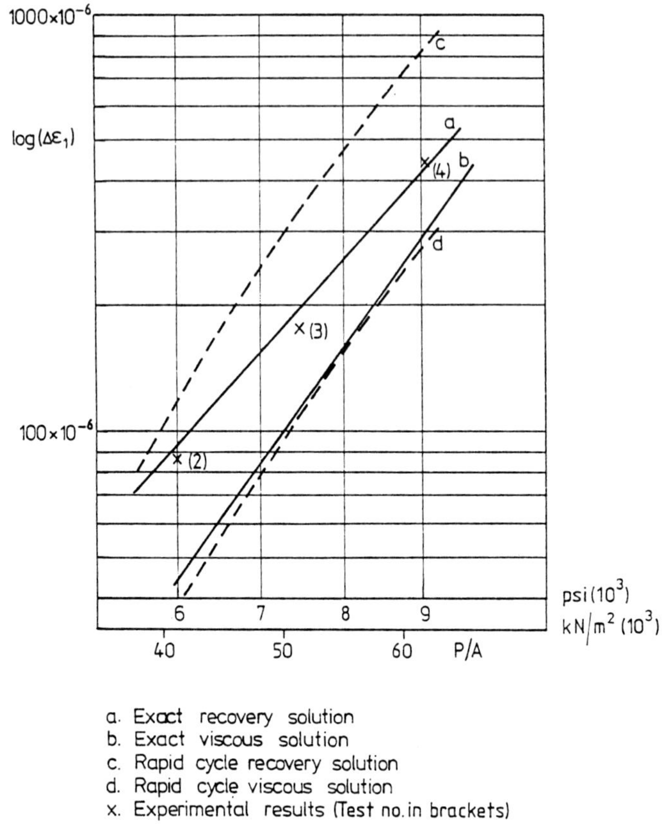

Fig. 4 Comparison between test results and various predictions for tests 2, 3 and 4.

given by Ponter and Megahed [6].

Prediction of rupture lives using the deformation stress histories and the Kachanov theory of rupture showed that copper generally has a longer creep rupture life than that predicted by the Kachanov theory modelled from constant stress data. Use of the rapid cycle deformation solution in conjunction with Kachanov theory to predict initial rupture gave an underestimate of rupture life which lay between 1/3 (test 6) and 1/1.6 (test 3). By comparison with predictions using the exact recovery model this underestimate was due, in about equal amounts, to the use of the rapid cycle solution in place of the exact solution and the deficiencies of the Kachanov model for variable temperature and stress.

Acknowledgements

The authors gratefully acknowledge the support of the Science Research Council for this work. The experiments were carried out with the assistance of Mr. C. Morrison and Mr. W. Topliss whose assistance is gratefully acknowledged.

References

1. Ponter, A.R.S. and Walter, M.H., "A Theoretical and Experimental Investigation of Creep Problems with Variable Temperatures", J. Applied Mechanics, Trans. ASME. Vol. 43, 1976, pp 639-644.

2. Mitra, S.K. and McLean, D., "Workhardening and Recovery in Creep", Proc. Roy. Soc., A295, 1966 pp. 2881295.

3. Ponter, A.R.S., "Deformation Bounds for the Bailey-Orowan Theory of Creep", Trans. ASME, Journal of Applied Mechanics, Vol. 42, Series E, 1975.

4. Ponter, A.R.S., "The Analysis of Cyclically Loaded Creeping Structures for Short Cycle Times", Int. J. Solids Structures Vol. 12, pp 809-825, 1976.

5. Ponter A.R.S. and Brown P.R., "The Finite Element Solutions of Rapid Cycling Creep Problems", Int. Jn. for Numerical Methods in Engineering, Vol. 12, pp 1001-1024, 1978.

6. Megahed, M.M. and Ponter, A.R.S., "Creep and Plastic Ratchetting in a Simple Structure", Department of Engineering, University of Leicester, Report No. 79-9, November 1979.

Une Formulation par la Methode des Elements Finis d'Analyse Dynamique Non-Lineaire Appliquée aux Systemes Tridimensionelles de Poutres et de Tuyaux

A. RICARD

Département TE.C - FRAMATOME

Tour FIAT PARIS LA DEFENSE

Sommaire :

Nous présentons une stratégie possible pour résoudre le probleme specifique du calcul de grands systèmes de tuyauteries en dynamique non linéaire. Les différents choix de résolution pour traiter les non linéarités géométriques et de comportement ainsi que celles dues à l'ovalisation des coudes de tuyauteries sont indiqués. Un exemple simple montre l'éfficacité de la méthode.

1. Introduction

L'analyse de structures en dynamique non linéaire est devenue, aujourd'hui, une nécessité pour le "design" de structures soumises à de fortes sollicitations. Cela est particulièrement vrai dans le domaine nucléaire. Les différents problèmes qui se posent alors à l'ingénieur sont de deux ordres : - un problème "technique" : peut-on prendre en compte toutes les non linéarités souhaitables ? et un problème "économique" : faire le calcul au moindre coût et pour cela utiliser une "bonne" modélisation.

Les fondements théoriques des calculs dynamiques non linéaires ont été établis depuis déja de nombreuses années (1) (2) (3) (4). Les non linéarités prises en compte sont en général les non linéarités de comportement (élasto-plasticité), et les non linéarités géométriques (grands déplacements...). Mais d'autres non linéarités peuvent se présenter lorsque l'on étudie des lignes de tuyauteries par exemple, ce sont principalement des conditions unilatérales (cas de supportages avec jeux ou impact sur d'autres structures) et pour les coudes de tuyauteries la simulation de l'ovalisation de la section des coudes.

Dans le cas de grands systèmes de tuyauteries ou de réseaux de poutres à étudier l'utilisation d'éléments tridimensionnels isoparamétriques du type pavé ou d'éléments de coques est exclus pour des raisons de coût, aussi doit-on utiliser des éléments tridimensionnels de type poutre permettant de tenir compte des non linéarités souhaitées. La modélisation est plus globale et ne permet pas en général de remonter à l'état de contraintes ou de déformations local en un point précis de la structure, par exemple en un point d'une section d'un tuyau donné. Si l'on désire une analyse plus fine de certaines zones il est possible d'utiliser pour cette partie seulement de la structure un maillage avec des éléments de coques.

2. Méthode proposée

Une analyse des différentes méthodes utilisées nous a conduit à faire le choix suivant d'algorythmes :

- pour le calcul dynamique on utilise la méthode implicite de NEWMARK avec $\alpha = 1/2$ et $\beta = 1/4$.
- pour le calcul en grands déplacements on utilise une formulation en Lagrangien réactualisé (utilisation de la matrice de rigidité géométrique K_σ et de forces correctives pour tenir compte des grandes rotations (5)).
- pour le calcul en élastoplasticité on utilise une formulation globale (6) où le critère est écrit directement en terme d'efforts et de moments.
- les conditions unilatérales sont résolues en modifiant la matrice de rigidité au moment de l'impact (7).
- l'ovalisation des coudes n'est prise en compte que de manière approchée par une modification des termes d'inertie (8). La méthode proposée par (6) dans le cas d'un calcul en petites déformations donne de bons résultats.

3. Limites de la méthode

Les principales limites sont relatives au calcul de tuyauteries et en particulier à la déformation des sections (ovalisation) des parties droites et au cours de l'impact. Aucune des solutions proposées dans la littérature me semble particulièrement performante. Aussi nous proposons de modéliser les parties du système de tuyauteries où ces phénomènes jouent un rôle important par un maillage de coques.

4. Algorythme de calcul

La méthode proposée est incrémentale (dans le temps, ou en statique en fonction de la charge) et itérative de manière à résoudre les équations d'équilibre sur le corps déformé à chaque instant. Le programme comporte la possibilité de faire les itérations de convergence nécessaires en grands déplacements et celles nécessaires pour la plasticité indépendamment les unes des autres ou de manière couplée. Cela dépend du problème traité et de l'importance relative des deux phénomènes.

L'organigramme général est de la forme suivante :

MAILLAGE

CALCUL INCREMENTAL

→ Pour chaque incrément

Réactualisation des coordonnées si nécessaire

Calcul des matrices de rigidité, de masse, d'amortissement.

Calcul des efforts appliqués.

→ CALCUL ITERATIF

→ Grands déplacements :

Calcul de la matrice $K\sigma$ et de forces correctrices et passage à l'itération de grands déplacements suivante

PLASTICITE :

→ Calcul des forces correctrices et passage à l'itération de plasticité suivante.

Couplage éventuel des deux types de calcul itératif

Passage à l'incrément suivant.

5. Exemple

Nous donnons à titre d'exemple le calcul dynamique en grands déplacements d'une poutre encastrée. La solution obtenue est très semblable à celle obtenue par la référence (3).

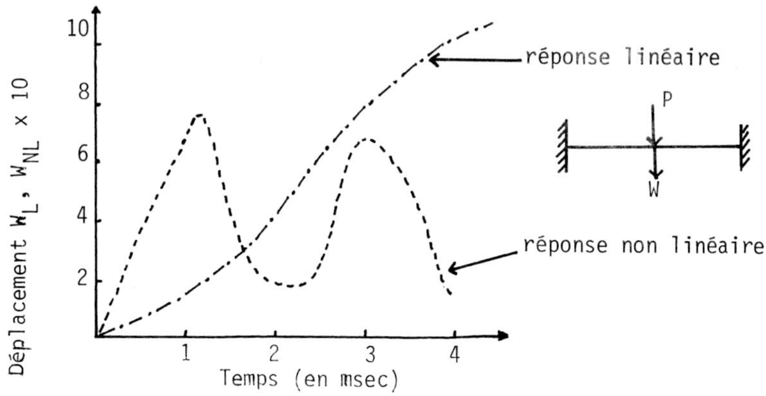

La charge P est une fonction échelon dans le temps P = 640.
Les caractéristiques du calcul sont les suivantes :
longueur : 10 $E = 3.\ 10^7$ $\nu = 0.$ masse volumique $\rho = 0{,}248\ 10^3$
$S = 0{,}125$ $I = 0{,}16276\ 10^3$. La poutre a été simulée par 10 éléments.
Le calcul dynamique en grands déplacements montre une réduction significative de la période et du déplacement maximum par rapport à un calcul analogue en petits déplacements. Pour un pas de temps donné, nous avons effectué 5 itérations de type grands déplacements et le temps CP sur CDC 7600 pour 40 pas de temps de 0,1 ms a été de 10 secondes.

6. Conclusion

L'introduction de cette méthode, dans un code général d'éléments finis, permet de traiter des problèmes de taille importante à un coût raisonnable. Elle ne nécessite pas l'emploi d'éléments très sophistiqués (comme des coques à double courbure ou des éléments spéciaux de coudes) et est donc d'un emploi industriel facile.

Références bibliographiques :

1 HIBITT, MARCAL, RICE : A finite élément formulation for problems of large strain and large displacement I.J.S.D. (1970) Vol 6 p 1069-1086

2 BREBBIA, CONNOR : Geometrically non linear finite element analysis A.S.C.E. E.M.2 (1969) p 463.483

3 MONDKAR, POWELL : Finite Element Analysis of non linear static and dynamic response I.J.N.M.E. Vol 11 (1977) p 499.520

4 ZIENCKIEWICZ : The finite element method in engineering science Mac Graw Hill (1971)

5 KOHNKE : Large deflection analysis of frame structures by fictitious forces I.J.N.M.E. Vol 12 (1978 p 1279 - 1294

6 HOFFMANN, ROCHE, MILLARD : Inelastic analysis of piping systems : A beam-type methode for creep and plasticity SMIRT Berlin (1976) L4/7.

7 WECAN : Manuel d'utilisation Rév. 0 Westinghouse (1978) RD center.

Prediction of Concrete Behaviour Using a Non-Linear Incremental Law

J.C. ROBINET - H. DI BENEDETTO

Ecole Nationale des Travaux Publics de l'Etat
Rue Maurice Audin - 69120 - VAULX-en-VELIN - FRANCE

Abstract

In this article we present a non-linear incremental law with an infinity of domains of expression (or a continuous domain) and to which a validity domain is associated. The hypotheses used for the description of the law are first presented. We are able to describe the rate independent part of the material with four hypotheses and only one more hypothesis allows us to take into account the viscous part. This new model in which the description of history takes an important part has been carried out in order to narrow down the observed differences between reality and calculation with classical laws to describe the cyclic phenomena.

A comparison between numerical integration of the proposed rheological model and experimental results over cyclical and monotonous path for concrete are presented in the last section.

1 - Introduction

Characterisation of a rheological law rests on fundamental principles [1] [2] (principle of objectivity, principle of determinism ...) which are dictated by the physical reality of the phenomenon they describe.

Deterministic principles applied in incremental form allow us to write $G\ (\delta\sigma,\ \delta\varepsilon,\ \delta t) = 0^{*}$ where δt is the time increment.

A first order linearisation of G [3] could be considered over a semi-straight support of $\delta\underset{\sim}{\sigma}$;

$$\delta\underset{\sim}{\varepsilon} = \underset{\sim}{M} \cdot \delta\underset{\sim}{\sigma} + \underset{\sim}{v}\ \delta t$$

where $\underset{\sim}{M} \cdot \delta\underset{\sim}{\sigma}$ represents the rate independent part, and $\underset{\sim}{v}\ \delta t$ the viscous part of the model.

In the following sections of this paper we only consider the rate independent part of the law : $\delta\underset{\sim}{\varepsilon} = \underset{\sim}{M} \cdot \delta\underset{\sim}{\sigma}$.

2 - Proposed law

The law that we propose rests on four hypotheses which are imposed by the fundamental principles and justified by the physical behavior of the material. Only one more hypothesis is necessary to describe the viscous part [3].

H.1 We consider that the non viscous matrix M is orthotropic when expressed in principal axes of the Cauchy-Euler strain tensor* ($\underset{\sim}{\varepsilon}_r$)

$$M_r = \begin{vmatrix} \frac{1}{E_1} & \frac{\nu_2^1}{E_2} & \frac{\nu_3^1}{E_3} & & & \\ \frac{\nu_1^2}{E_1} & \frac{1}{E_2} & \frac{\nu_3^2}{E_3} & & 0 & \\ \frac{\nu_1^3}{E_1} & \frac{\nu_2^3}{E_2} & \frac{1}{E_3} & & & \\ & & & 2G & & \\ & 0 & & & 2H & \\ & & & & & 2I \end{vmatrix}$$

In this set of axes the non viscous part of the law takes the following form : $\delta\underset{\sim}{\varepsilon}_{Ar} = M_r \; \delta\underset{\sim}{\sigma}_r$

H.2 We would assume that the stress paths simulate the memory of the material if they lead to the same state of stress and the same specific volume. This important hypothesis is used to find a simple sollicitation path we call "elementary path" [4] which leads to a material having the same response as the real one.

H.3 This hypothesis allows us to determine the functions E_i and ν_i^j, which are necessary for the description of M_r. In this study we have only kept a linear variation with θ_i [4] ($\theta_i = (\delta\vec{\sigma}, I)$, I is an orthotropy direction).

H.4 The preceding three hypotheses do not permit to define the fonctions G, H and I which determine the distorsion of the material. As pointed out by experimental studies performed on rocks, the inverse of the Young modulus follows an ellipsoïdal law, once the coordinate axes of the expression of the law rotate. These experimental results constitute the fourth hypothesis. After calculation [4] we can get the G, H and I functions.

The reader is in a position to establish through these four hypotheses that the proposed law interprets a continuous evolution of the response of the material as a function of the stress tensor path "orientation".

---

* If F is the linear tangent transformation

$\underset{\sim}{\varepsilon}_r = \tilde{F}^{-1} \, . \, F^{-1} = 1 - 2\,\underset{\sim}{\varepsilon}_A$; $\underset{\sim}{\varepsilon}_A$ Almansi strain tensor

3 - Validity domain of the rheological law

The following development allows us to determine an incremental limit for the law.

The theoretical interpretation of the appearance of discontinuity in the material [3], [4] defines the instability criterion of the model

$$\det\,(n\underset{\sim}{C}n) = 0$$

This criterion is used to define the beginning of cracks in concrete

4 - Application of the rheological law

The law whose parameters have been found over the elementary path [4] can be "numerically integrated". We present in this paragraph some comparison of numerical integrations and experiments carried out on the concrete tested by Newman [5].

Simulation of a deviatoric triaxial test

Figure 1 shows the results of the numerical integration over a deviatoric triaxial path of revolution. The dashed curves represent the instability domain (probability of crack) of the rheological law.

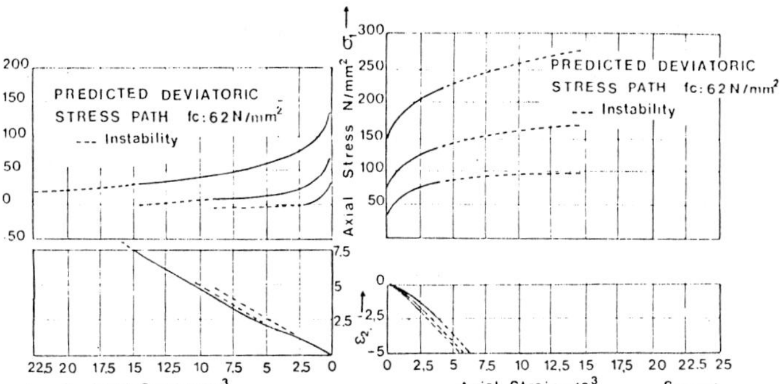

Figure 1

Comparisons of the direction increment vectors of theoretical and experimental deformation at rupture over the triaxial path of revolution (figure 2).

Validity domain of the rheological law

For a material considered initially as homogeneous we have searched for the possibility of the appearance of fissures at incremental level over different stress-paths. Results of this research are shown, figure 2 and 3.

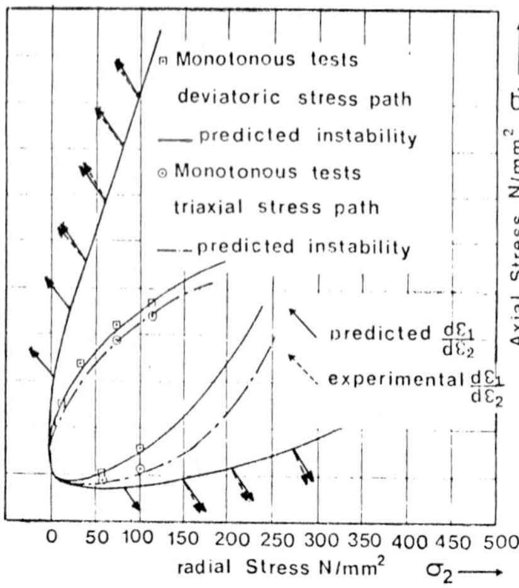

Figure 2

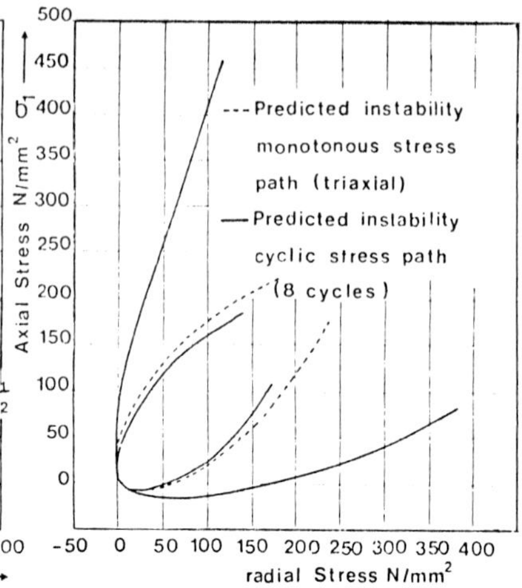

Figure 3

Evolution of the validity domain with the number of cycles (figure 3).

5 - Conclusion

We have presented in this paper a non linear rheological incremental law with a continuous domain of expression for concrete.

The hypotheses developped result in a law which yields an output depending on the "direction" of the stress path, on the past history of the material, as well as on the appearance of cracks.

Worked numerical examples illustrate a good agreement between observed and calculated behaviour, and prove the potentialities of the proposed model for practical computations with concrete.

Slort references listing

1 Mandel : Cours de Mécanique des Mileux Continus - Tome 1 Gauthier-Villars - Paris (1966)

2 C. Truesdell and W. Noll : The non-linear field theories - vol. III.3 Handbook of physics - Springer - (1965)

3 J.C. Robinet ; H. Di Benedetto : Présentation d'une loi rhéologique non linéaire en écriture incrémentale pour les sols, et comparaison avec les différents modèles existants - Cahier du Groupe Français de Rhéologie (1980)

4 J.C. Robinet ; H. Di Benedetto : Loi rhéologique en écriture incrémentale pour les sols - Journée Rhéologie - E.N.T.P.E. (1979)

5 M.D. Kotsovos ; S.B. Newman : Behavior of concrete under multiaxial stress ACI - Journal Proceedings - vol. 74 - (1977)

Description of Anisotropic Damage Application to Elasticity

F. SIDOROFF

Laboratoire de Mécanique des Solides

Ecole Centrale de Lyon - 69130 ECULLY - France

Summary

A general approach to anisotropic damage is presented starting from the concept of effective stress tensor. Some existing possibilities are discussed and the special case of uncoupled anisotropic damage is presented. The influence of damage on the elastic response is then investigated and it is shown that direct substitution of the effective stress tensor in the constitutive equation usually is not admissible. Some possibilities are presented to overcome this difficulty.

1. Introduction

The concept of damage was introduced by Kachanov in 1958 for the description of tertiary creep, and has been widely used since then for life-time predictions or in cases when internal degradation becomes important and cannot be ignored in the description of material behaviour. The one dimensional case is well understood [1] as well as the three dimensional extension to isotropic damage [2]. However, there is strong experimental evidence [3],[4] that damage initiation and growth, essentially is an anisotropic phenomenon and it is important to develop a consistent theory for anisotropic damage.

Instead of the usual scalar damage parameter D, anisotropic damage requires the introduction of a damage tensor, but many possibilities are left. Furthermore some fundamental difficulties arise in the development of the theory, specially in connection with thermodynamics. This paper is devoted to a general approach to anisotropic damage. Some of the existing works will be reviewed and the above mentioned difficulties will be pointed out. We shall be mainly concerned with the influence of damage on the elastic response of the material. This is only one special aspect of anisotropic damage but it is a very good example for our purpose.

2. The effective stress tensor

The damage process results in a weakening of the material and leads to the failure of the structure. This process is usually described by the introduction of a damage variable d which can be viewed as a macroscopic measure of the internal degradation state of the material. The mechanical response of the material gives the strain tensor at time t as a functional of the stress and damage histories up to time t.

$$\underline{\varepsilon}(t) = \mathcal{E}\{\underline{\sigma}^t, d^t\} = \tilde{\mathcal{E}}\{\underline{\sigma}^t\} \tag{1}$$

Of course, in a mechanical theory, the damage process is governed by the sollicitation; this allows the elimination of the damage history in (1), resulting in a damaged functional $\tilde{\mathcal{E}}$. The undamaged response, which is the response of an ideal undamaged material, is obtained when there is no damage and is assumed to be known

$$\mathcal{E}\{\underline{\sigma}^t, 0^t\} = \mathcal{E}_0\{\underline{\sigma}^t\} \tag{2}$$

At the basis for all present developments, lies Kachanov's idea of effective stress. In the one dimensional or three dimensional isotropic case (where the damage variable d reduces to a scalar variable D), it may be expressed in the following way:

Basic postulate. The damaged response of the material is obtained by replacement of the stress tensor $\underline{\sigma}$ by an "effective stress tensor" $\underline{\tilde{\sigma}}$ in the undamaged response

$$\underline{\varepsilon}(t) = \mathcal{E}\{\underline{\sigma}^t, D^t\} = \mathcal{E}_0(\underline{\tilde{\sigma}}^t) \tag{3}$$

$$\underline{\tilde{\sigma}}(t) = \frac{\sigma(t)}{1 - D(t)} \tag{4}$$

Thus the theory only requires formulation of the evolution law for D, failure occuring when D reaches some critical value.

3. Anisotropic damage

Extension to anisotropic damage is quite straightforward since the basic postulate (3) can remain unchanged while the definition (4) of the effective stress tensor should be replaced by some more general transformation law, accounting for anisotropy, and which quite generally can be written as

$$\underline{\tilde{\sigma}} = \underline{\underline{M}}(d)\ \underline{\sigma} \tag{5}$$

where $\underline{\underline{M}}(d)$ is a linear operator on second order symmetric tensors (i. e. a fourth order tensor) and where the damage variable d now must be of tensorial character. This approach raises two problems and one question. The two problems are: 1. What kind of tensor should be introduced to describe damage ? 2. Which transformation law (5) should be used ? Many answers have been proposed and we shall present them thereafter. The question is: Does this approach lead to a consistent formulation of the damaged response ? We shall see that the answer, at least in the simplest case of the elastic response, can be "no", and this may be one of the most fundamental problems about anisotropic damage.

The simplest possibility for an anisotropic damage variable is a second order symmetric tensor $\underline{D}$. This choice implies some limitations, but it has been widely used [5], [6], [7]. The transformation law is then chosen as a simple generalization of (4) (see [7], § 2.3), for instance [5],

$$\underline{\tilde{\sigma}} = \frac{1}{2}\{(\underline{1} - \underline{D})^{-1}\ \underline{\sigma} + \underline{\sigma}\ (\underline{1} - \underline{D})^{-1}\} \tag{6}$$

Another approach is a direct identification of damage with the operator $\underline{\underline{M}}$ in (5), [8]. It results in a fourth order damage tensor $\underline{\underline{D}}$ and the following transformation law

$$\underline{\tilde{\sigma}} = (\underline{\underline{1}} - \underline{\underline{D}})^{-1}\ \underline{\sigma} \tag{7}$$

Starting from the elastic case, an eighth order damage tensor has also been introduced [9].

Other theories are expressed in terms of principal stresses [4] and the transformation law (5) is written, without referring explicitly to any tensor representation, as

$$\tilde{\sigma}_1 = \frac{\sigma_1}{1 - D_1}\ ,\quad \tilde{\sigma}_2 = \frac{\sigma_2}{1 - D_2}\ ,\quad \tilde{\sigma}_3 = \frac{\sigma_3}{1 - D_3} \tag{8}$$

with three principal damages D_1, D_2, D_3. This is a reasonable assumption if the material is subjected to a loading history with fixed principal axes. Indeed, in this case, the principal axes of $\underline{\sigma}$ and $\underline{\tilde{\sigma}}$ coincide and the transformation law (4) can be written as

$$\tilde{\sigma}_1 = \Delta_{11}\sigma_1 + \Delta_{12}\sigma_2 + \Delta_{13}\sigma_3 \quad , \quad \tilde{\sigma}_2 = \ldots \tag{9}$$

Equation (8) is obtained if the non diagonal terms Δ_{12}, Δ_{13}, ... vanish, i.e. if there is no coupling between the principal stresses. If a damage description is such that it reduces to (8) for a loading with fixed principal axes, it will be called "uncoupled anisotropic damage". This is in particular the case if $\underline{D}$ is a second order tensor and if $\underline{\tilde{\sigma}}$ is given by (6) or some other similar expression [7], but it can also occur for other descriptions.

4. Damaged elastic response

Application of the basic postulate of § 2 to the elastic response of an initially isotropic material is quite straightforward. The damaged and undamaged elastic responses are given by

$$\underline{\varepsilon} = \tilde{\underline{\underline{A}}}(d)\ \underline{\sigma} \tag{10}$$

$$\underline{\varepsilon} = \underline{\underline{A}}_\circ\ \underline{\sigma} = \frac{1+\nu}{E}\underline{\sigma} - \frac{\nu}{E}\operatorname{tr}\underline{\sigma}\ \underline{1} \tag{11}$$

It follows from thermodynamic considerations or from homogeneization results [7] that the damage elastic response is symmetric (as an operator on symmetric tensors). The basic postulate, with the help of (11) and (5) gives

$$\tilde{\underline{\underline{A}}}(d) = \underline{\underline{A}}_\circ\ \underline{\underline{M}}(d) \tag{12}$$

Unfortunately the symmetry condition for $\tilde{\underline{\underline{A}}}(d)$ implies very strong restrictions on the transformation law $\underline{\underline{M}}$.

As a first result, we shall show that uncoupled anisotropic damage, as defined in § 2, is not compatible with the basic postulate. Indeed, inserting (8) in (11), we obtain

$$\begin{bmatrix}\varepsilon_1\\ \varepsilon_2\\ \varepsilon_3\end{bmatrix} = \frac{1}{E}\begin{bmatrix}\frac{1}{1-D_1} & -\frac{\nu}{1-D_2} & -\frac{\nu}{1-D_3}\\ -\frac{\nu}{1-D_1} & \frac{1}{1-D_2} & -\frac{\nu}{1-D_3}\\ -\frac{\nu}{1-D_1} & -\frac{\nu}{1-D_2} & \frac{1}{1-D_3}\end{bmatrix}\begin{bmatrix}\sigma_1\\ \sigma_2\\ \sigma_3\end{bmatrix} \tag{13}$$

which is not symmetric, unless $D_1=D_2=D_3$, i.e. unless damage is isotropic.

More generally, the symmetry condition of $\tilde{\underline{\underline{A}}}(d)$, implies restrictions on $\underline{\underline{M}}(d)$. In particular, if the damage transformation law $\underline{\underline{M}}$ is assumed to be universal, i.e. to be the same for all materials, as is (4), then it can be shown that it is so restricted that if the stress tensor $\underline{\sigma}$ is isotropic (or deviatoric) so is the effective stress tensor $\underline{\tilde{\sigma}}$, which seems contrary to the concept of anisotropic damage.

Thus it results from the preceding analysis that, at least for the elastic response, we must choose between the basic postulate and a simple description of damage, since these two ideas are not compatible. In the following, these two possibilities will be exemplified.

5. Energetic identification

The first approach consists in keeping a simple description of damage, (6) or a similar one, and rejecting the basic postulate, i.e. a direct identification of the constitutive equation of the damaged material from the undamaged. Instead it seems appropriate to use an energetic identification, [7]. The constitutive equation (10) and (11) can be written in terms of the damaged and undamaged complementary energy

$$\underline{\varepsilon} = \partial\tilde{V}/\partial\underline{\sigma} \quad , \quad \tilde{V}(\underline{\sigma},d) = \frac{1}{2}\,\underline{\sigma}\,\tilde{\underline{\underline{A}}}(d)\,\underline{\sigma} \tag{14}$$

$$\underline{\varepsilon} = \partial V_\circ/\partial\underline{\sigma} \quad , \quad V_\circ(\underline{\sigma}) = \frac{1+\nu}{2\,E}\,(\mathrm{tr}\,\underline{\sigma}^2) - \frac{\nu}{2E}\,(\mathrm{tr}\,\underline{\sigma})^2 \tag{15}$$

We shall assume that the damaged complementary energy is obtained by substituting the stress by the effective stress in the undamaged complementary energy

$$\tilde{V}(\underline{\sigma},d) = V_\circ(\underline{\tilde{\sigma}}) \tag{16}$$

From this follows the damaged elastic law as

$$\tilde{\underline{\underline{A}}}(d) = \underline{\underline{M}}(d)^T\,\underline{\underline{A}}_\circ\,\underline{\underline{M}}(d) \tag{17}$$

Within this framework, an analysis of the damaged induced anisotropy (with (6) and similar transformation laws) has been performed [7].

This approach is simple and natural. Furthermore, since the damage transformation law is not restricted, it is compatible with the existing creep damage theories [4], [5]. However it leaves a great deal of freedom for the description of other phenomena than elasticity, because the guideline of the basic postulate disappears. In particular, the extension to plasticity is now being investigated.

6. Material dependent damage law

The alternative approach consists in trying to find a damage transformation law allowing us to keep to the basic postulate. However, in this case, it is necessary to choose it dependent on the material, i. e. on Poisson's ratio ν, since, as mentioned in § 4, a universal damage transformation law was too restricted.

A first example has been developed by Chaboche [8] starting from a fourth order damage tensor (7). It is based on some homogeneization results on elastic materials weakened by arrays of cracks. General anisotropic damage is obtained as the combination of isotropic damage with "perfect anisotropic damage" corresponding to cracks perpendicular to the greatest principal stress direction. The model of an eighth order damage tensor [9] also falls within this framework.

Another example can be obtained with a symmetric second order damage tensor $\underline{D}$. Starting from tensor representations and expressing the symmetry condition for the damaged elastic response as defined by (12), we can obtain admissible transformation laws. For instance if we introduce, instead of (6),

$$\underline{\tilde{\sigma}} = \frac{1}{2}\{(\underline{1} - \underline{D})^{-1}\,\underline{\sigma} + \underline{\sigma}\,(\underline{1} - \underline{D})^{-1}\} + \frac{\nu}{1-2\nu}\,\mathrm{tr}\{\underline{\sigma}\underline{D}(\underline{1} - \underline{D})^{-1}\}\,\underline{1}$$

then the symmetry condition is satisfied.

This approach keeps to the original introduction of damage, however it is rather complicated and essentially relies on elasticity which may be is not the essential phenomenon. It is therefore essentially disconnected from other damage theories.

7. Conclusion

We have presented some of the problems related to the description of anisotropic damage as well as some ways of overcoming them. The choice between the various approaches presented above is not yet clear. In the elastic case experimental results and homogeneization could clarify the situation. Indeed in this simplest case, the influence of damage can be directly observed or easily simulated, 7 . For this reason, this aspect of damage, even if of little practical interest, is worth studying for the understanding and development of the conceptual machinery.

Extension of these concepts to other aspects of damage - plastic damage, creep, fatigue - will require two kinds of relationships: constitutive equations and evolution law for damage. Constitutive equations deal with the influence of damage on the material behaviour and the coupling between damage and other aspects of mechanical behaviour. The basic approach will be some kind of substitution of the effective stress tensor $\underline{\tilde{\sigma}}$ in place of the stress tensor $\underline{\sigma}$. However, once the basic postulate rejected, this substitution can be achieved in many different ways. Anyway thermodynamic consistency should be kept in mind. The guideline of the one dimensional theory 1 is useful but uneasy because - except for creep which may be simpler - it is not usually expressed in thermodynamic terms.

The evolution law for damage involves some other kind of problems. Non linear thermodynamics of irreversible processes could be useful, since it tells us that the rate of damage essentially depends on the thermodynamic force G associated to the damage variable d and which is defined as

$$\Phi = \ldots + G\,\dot{d} > 0 \qquad , \qquad G = -\frac{\partial W}{\partial d} \tag{18}$$

where W is the energy and Φ the dissipation. Another restriction comes from the fact that material degradation cannot be reversed (at least in an isothermal theory excluding recovery) and therefore that d must be, in some sense, an increasing function of time. This implies restrictions both on the evolution law and on G which according to (18) must in some sense be positive. For instance if damage is described by a symmetric tensor $\underline{D}$, it is reasonable to assume that its principal values are increasing functions of time and it can be shown that this is ensured if the time derivative $\underline{\dot{D}}$ is a positive definite tensor, which is a natural condition. But for other descriptions of damage, this may be not so easy.

1 LEMAITRE J. - CHABOCHE J.L.: Aspect phénoménologique de la rupture par endommagement. J. Mécanique Appliquée, vol.2, n°3, 1978, p.317 -365.

2 LEMAITRE J.: Damage modelling for prediction of plastic or creep fatigue-failure in structures. SMIRT 5, August 1979.

3 HAYHURST D.R.: Creep rupture under multiaxial states of stress. J. Mech. Phys. Solids, vol.20, 1972, p.381-390.

4 HAYHURST D.R. - LECKIE F.A.: The effect of creep constitutive and damage relationships upon the rupture time of a solid circular torsion bar. J. Mech. Phys. Solids, vol.21, 1973, p.431-446.

5 MURAKAMI S. - OHNO N.: A constitutive equation of creep damage in polycrystalline metals. Euromech Colloquium 111, Marianske Lazne, September 1978.

6 MURZEWSKI J.: Cumulative damage of solids for random stress. Engng Fracture Mech., vol.8, 1976.

7 CORDEBOIS J.P. - SIDOROFF F.: Damage induced elastic anisotropy. Colloque Euromech 115, Villard de Lans, Juin 1979.

8 CHABOCHE J.L.: Le concept de contrainte effective appliqué à l'élasticité et à la viscoplasticité en présence d'un endommagement anisotrope. Colloque Euromech 115, Villard de Lans, Juin 1979.

9 CHABOCHE J.L.: Description thermodynamique et phénoménologique de la viscoplasticité cyclique avec endommagement. Thèse, Université P.et M.Curie, 1978.

Mixed Work-Hardening Law for Steels under Complex Loading Histories and its Application to Finite Element Analysis

J. SUHARA

Deparment of Naval Architecture
Kyushu University
Fukuoka, JAPAN

1. Introduction

While the behavior of material under monotonic loading is made traceable in the light of existing theory of plasticity, the cases of cyclic loading that varies irregularly are hard to describe faithfully by the same theory. To improve this situation, a method to simulate the stress-strain behavior of steel under uniaxial loading based on computer is firstly proposed. Generalization of this simulation to three dimensional case by means of yield function for mixed hardening law is proposed. A finite element analysis of a cracked plate shows validity of describing the material behavior through this approach.

2. Simulation of Uniaxial Stress-Strain Behavior of Steel

To simulate numerically the stress-strain behavior of steel under regular or irregular loading exceeding elastic limit, the material constants and the associated experimental laws governing the plastic internal variables are determined previously through the tests carried out using several test specimens made of the same material. The laws of same type are found generally in mild steels and high strength steels. The following formulae are shown for a mild steel which is previously annealed (carbon content : 0.12 %, yield point : 27.0 kg/mm^2, tensile strength : 43.9 kg/mm^2, modulus of elasticity : 2.17×10^4 kg/mm^2), and they are expressed by true stress in kg/mm^2 and true strain in mm/mm. Nomenclature is illustrated in Fig. 1 and Fig. 2. For monotonic loading, stress σ is expressed by total strain ε, i.e.

$$\sigma = \begin{cases} \mathrm{sign}(\varepsilon)\cdot 2.17 \times 10^4 \cdot \varepsilon, & (0 \leqq |\varepsilon| < 1.244 \times 10^{-3}) \\ \mathrm{sign}(\varepsilon)\cdot 27.0, & (1.2244 \times 10^{-3} \leqq |\varepsilon| < 0.0198), \\ \mathrm{sign}(\varepsilon)\cdot\{63.8 - 3.38 (\varepsilon + 0.072)^{-1}\}, & (|\varepsilon| \geqq 0.0198) \end{cases} \quad (1)$$

Denoting the maximum plastic strain by ε_p and the strain amplitude by

$\Delta\varepsilon_B$, for the first cycle of equal stress amplitude (see Fig. 1), we have

$$\Delta\varepsilon_B = \begin{cases} 0.0198 - \varepsilon_p, & (\varepsilon_p < 0.0117), \\ 0.0022 + 0.51\ \varepsilon_p, & (\varepsilon_p \geqq 0.0117). \end{cases} \tag{2}$$

$$d_1/\sigma_1 = 12.03 \cdot \varepsilon_p^{1.08} \cdot \exp (-39.2\ \varepsilon_p). \tag{3}$$

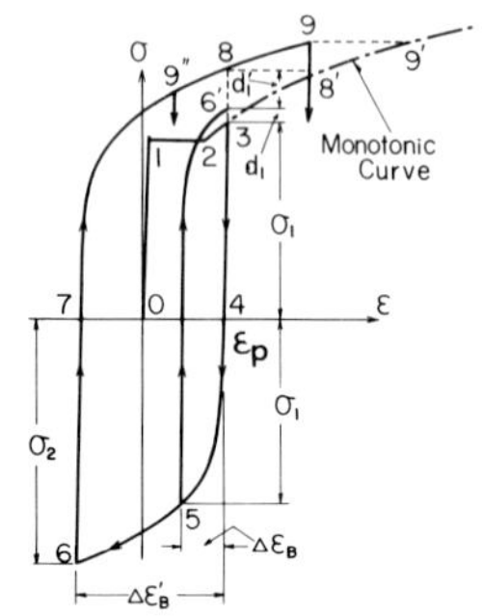

Fig. 1. Typical first stress-strain cycles.

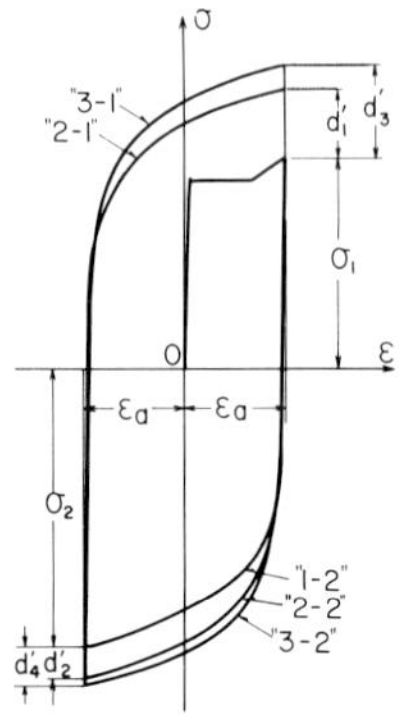

Fig. 2. Stress-strain cycles of equal strain amplitude.

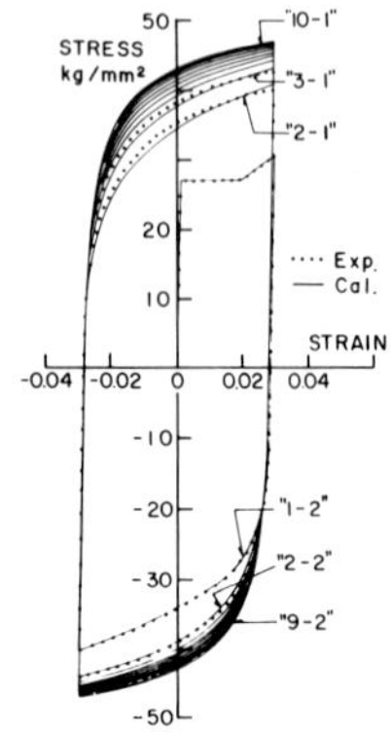

Fig. 3. Comparison of simulation and experimental data.

For the cycle of unequal stress amplitude (see Fig. 1), we have

$$d_1'/d_1 = \begin{cases} 1.04\ \{(\Delta\varepsilon_B'/\Delta\varepsilon_B) + 0.1\ \}^2 - 0.26, & (0 < \Delta\varepsilon_B'/\Delta\varepsilon_B \leqq 1), \\ 1.13\ (\Delta\varepsilon_B'/\Delta\varepsilon_B) - 0.13. & (1 < \Delta\varepsilon_B'/\Delta\varepsilon_B) \end{cases} \tag{4}$$

Denoting the stress increment due to the cycle of equal strain amplitude ε_a by d_M' (M = 1,2,....) (see Fig. 2), we have

$$\begin{aligned} d_M'/d_1' &= 0.452 \log_{10} M + 1.02, \quad (M = 1,3,.....) \\ d_M'/d_2' &= 0.648 \log_{10} M + 1.01. \quad (M = 2,4,.....) \end{aligned} \tag{5}$$

$$(d_2'/\sigma_2)/(d_1'/\sigma_1) \doteqdot \text{constant} = 0.28. \tag{6}$$

We assume that the stress varies linearly with the strain in elastic range and the hardening curves are described by Ramberg-Osgood type equations within the strain range $\Delta\varepsilon_B$ and the repeating cycle range of equal strain amplitude ε_a. For the process beyond those ranges, the hardening curves are assumed to be continued on them and extend as the corresponding part of monotonic curve (curve 89 is idendical with curve 8'9' in Fig. 1, for example). By employing further formulae (1) to (6) in accordance with the variation of stress or strain, numerical simulation can be realized [1]. The examples for comparison between the simulations and the experiments are shown in Fig. 3 and Fig. 4, where good agreement is seen between them.

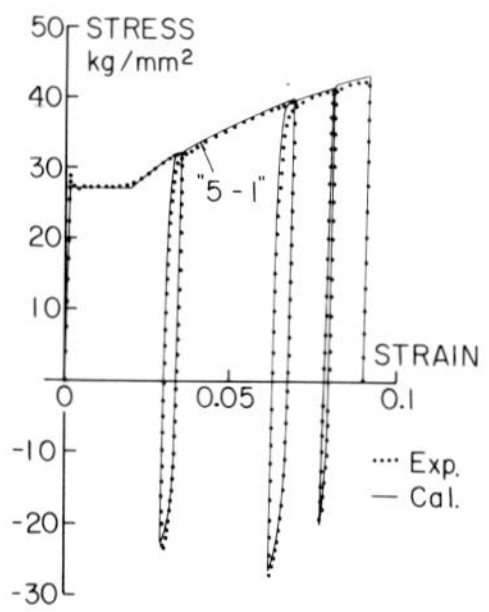

Fig. 4. Stress-strain simulation under arbitrarily varied loading.

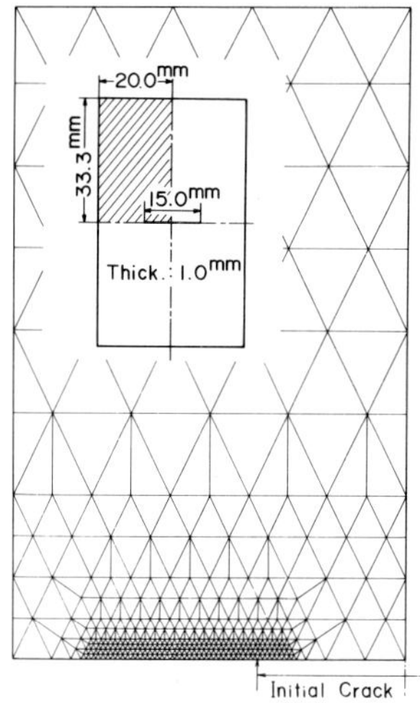

Fig. 5. Geometry of cracked plate and finite element mesh.

3. Mixed Work Hardening Law and Its Application

Simplified mixed work hardening law proposed by Tanaka et al. [2], where expansion, Ziegler type translation and distortion of yield surface are considered , and where a scaler parameter assumed as a function of equivalent plastic strain is used to control the effect of quadratic terms of plastic strain components in the yield function, is improved in this study as follows : (1) Incorporation with the uniaxial simulation obtained through the formulation mentioned above, (2) Implementation of the correction of yield function at initial yielding stage, taking yielding plateau and succeeding hardening effects into account, (3) Correction of consistency condition due to the distortion of the yield surface to conform to the generalization of uniaxial simulation, (4) Introduction of the effect of partially reversed loading which may appear in the case of multiaxial reloading, depending on the directions of normal vectors at the preceding and subsequent yield points on a yield surface.

The law mentioned above is applied to a finite elment analysis, where a cracked steel plate is subjected to repeating and uniformly distributed load at ends. The geometry and finite element mesh are shown in Fig. 5. Obtained stress-strain cycles of an element near initial crack tip and the illustation of the assumed schemes of loading and artificial crack propagation, comparing the results based on the mixed work hardening law and Ziegler's kinematic hardening law are shown in Fig. 6. Fig. 7 shows the prediction of crack growth of the plate obtained by integrating Paris' relation in which the values of ranges of stress intensity factor obtained through this finite element analysis were used, taking the effect of crack closure [3] into account as boundary condition of crack surface. The effect of fatigue crack retardation due to the application of single peak overload between regularly repeating load is clearly observed in these results, and the definite discrepancies between the results based on the respective laws are recognized.

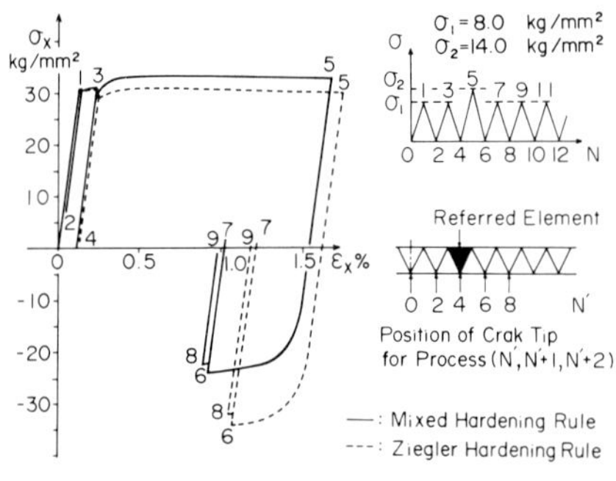

Fig. 6. Longitudinal stress-strain cycles in an element.

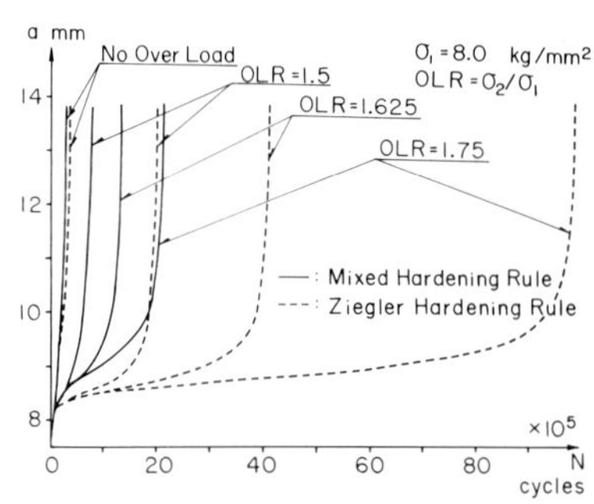

Fig. 7. Predicted half crack length v.s. actual number of load cycles.

4. Conclusion

It is concluded that the faithfulness of the constitutive law has remarkable influence on the results of elastic-plastic analysis, when irregular variation of load and Baushinger effect are predominant.

Refernces

1. Suhara, J.; et al.: The stress-strain behaviors of mild steels under complex loading history. Rep. Res. Inst. App. Mech. Kyushu Univ., XXXIII, 72 (1975) 1-30.

2. Tanaka, M.; et al.: On generalized kinematic hardening theory of plasticity. Ing. Arch. 44 (1975) 255-568.

3. Elber, W.: Fatigue crack closure under cyclic tension. Eng. Frac. Mech., 2 (1970) 37-45.

Incremental Analysis of Elasto-Plastic Shells of Revolution and its Application to Practical Design

M. TANAKA and M. HAMADA

Department of Mechanical Engineering
Faculty of Engineering
Osaka University
Yamada-Kami, Suita 565, JAPAN

Summary

An incremental formulation is presented for the elastoplastic problems of shells of revolution in which the effects of large deformation are taken into account. Some simplified theories of plasticity are incorporated into the formulation. A solution procedure is proposed in which the finite difference scheme is used together with the Fourier series expansion. It is shown that the elastoplastic computation is applicable to the lifetime estimation for the low-cycle fatigue of some axisymmetric shells.

1. Introduction

In accordance with the rapid development of high-speed electronic computers, it has been possible to analyze with sufficient accuracy the linear as well as the nonlinear problems of any structure by means of computer-oriented methods of solution. As for the thin shell problems, the solution procedures for linear elastic problems were first explored and then those for nonlinear problems are at present being investigated extensively. In references [1] and [2], for example, we can find a number of such investigations.

The authors have been so far concerned with the elastoplastic problems of shells of revolution and established a solution procedure based on the finite difference approximation[3-6]. In the present report, we show a unified approach to the thin shell problems including both geometrical and physical nonlinearities. The constitutive relations of a combined hardening model are applied to the formulation. The low-cycle fatigue of bellows-type expansion joints is investigated by using the elastoplastic analysis and the appropriateness of the procedure is ascertained by comparison with experiment. Finally, a design formula is proposed for the low-cycle fatigue of the bellows.

2. Incremental Formulation

2.1 Basic Equations

The coordinate system and notations of the shell of revolution is shown in Fig.1. The middle surface of the shell is expressed by

$$r=r(s) \tag{1}$$

Here and henceforth we use the following notations:

$$\xi=s/a,\quad \rho=r/a,\omega_{\xi}=a/R_{s},\omega_{\theta}=a/R_{\theta},(\ )'=d(\ )/d\xi,\gamma=\rho'/\rho \tag{2}$$

where a is the reference length of the shell. An arbitrary point of the shell can be expressed in the orthogonal coordinate system-$\xi\theta\zeta$ as shown in Fig.1.

We define the resulting forces, resulting moments and external loads acting on an infinitesimally small element of the middle surface as shown in Fig.2. The displacements and rotations of a point on the middle surface are shown in Fig.3. Denoting by Δ an increment, we can express the basic equations except the constitutive relations of Sanders' nonlinear theory of elastic shells[7] as follows:

The relations between membrane strains and displacements;

$$\begin{aligned}
\Delta\varepsilon_{\xi m}&=(1/a)(\Delta U_{\xi,\xi}+\omega_{\xi}\Delta W)+\Delta H_1\\
\Delta\varepsilon_{\theta m}&=(1/a)[(1/\rho)\Delta U_{\theta,\theta}+\gamma\Delta U_{\xi}+\omega_{\theta}\Delta W]+\Delta H_2\\
\Delta\gamma_{\xi\theta m}&=(1/a)[\Delta U_{\theta,\xi}+(1/\rho)\Delta U_{\xi,\theta}-\gamma\Delta U_{\theta}]+\Delta H_3
\end{aligned} \tag{3}$$

where

$$\Delta H_1=\Delta(\phi_{\xi}^2+\phi^2)/2,\quad \Delta H_2=\Delta(\phi_{\theta}^2+\phi^2)/2,\quad \Delta H_3=\Delta(\phi_{\xi}\phi_{\theta}) \tag{4}$$

The relations between curvature changes and rotations;

$$\begin{aligned}
\Delta\kappa_{\xi}&=(1/a)\Delta\phi_{\xi,\xi}\\
\Delta\kappa_{\theta}&=(1/a)[(1/\rho)\Delta\phi_{\theta,\theta}+\gamma\Delta\phi_{\xi}]\\
\Delta\kappa_{\xi\theta}&=(1/a)[\Delta\phi_{\theta,\xi}+(1/\rho)\Delta\phi_{\xi,\theta}-\gamma\Delta\phi_{\xi}]+(\omega_{\theta}-\omega_{\xi})\Delta\phi/a
\end{aligned} \tag{5}$$

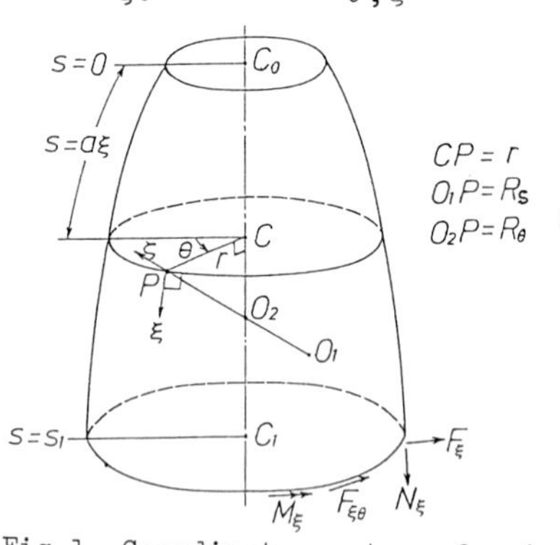

Fig.1 Coordinate system of sxi-symmetric shell

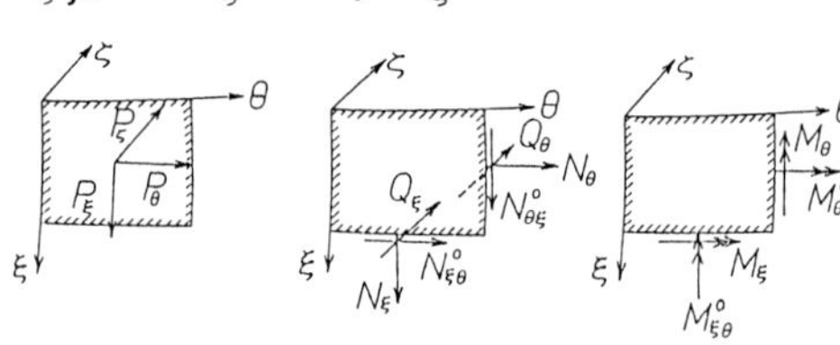

Fig.2 Loads, forces and moments

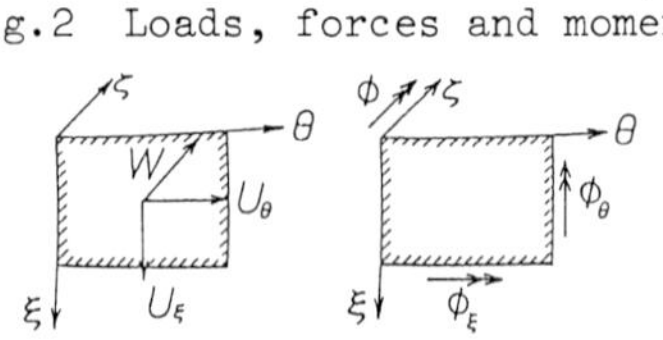

Fig.3 Displacements and rotations

The relations between rotations and displacements;

$$\Delta\phi_\xi=(1/a)(-\Delta W_{,\xi}+\omega_\xi\Delta U_\xi)$$

$$\Delta\phi_\theta=(1/a)[-(1/\rho)\Delta W_{,\theta}+\omega_\theta\Delta U_\theta] \tag{6}$$

$$\Delta\phi=(1/2a)[\Delta U_{\theta,\xi}+\gamma\Delta U_\theta-(1/\rho)\Delta U_{\xi,\theta}]$$

The equilibrium equations;

$$a[\rho\Delta N_{\xi,\xi}+\Delta N_{\xi\theta,\theta}+\rho'(\Delta N_\xi-\Delta N_\theta)]+\rho\omega_\xi\Delta M_{\xi,\xi}+(3\omega_\xi-\omega_\theta)\Delta M_{\xi\theta,\theta}/2$$
$$+\rho'\omega_\xi(\Delta M_\xi-\Delta M_\theta)+a^2\rho\Delta P_\xi=\Delta H_4 \tag{7a}$$

$$\rho\Delta M_{\xi,\xi\xi}+2\rho'\Delta M_{\xi,\xi}-\rho\omega_\xi\omega_\theta\Delta M_\xi+(1/\rho)\Delta M_{\theta,\theta\theta}-\rho'\Delta M_{\theta,\xi}+\rho\omega_\xi\omega_\theta\Delta M_\theta$$
$$+2(\Delta M_{\xi\theta,\xi\theta}+\gamma\Delta M_{\xi\theta,\theta})-a\rho(\omega_\xi\Delta N_\xi+\omega_\theta\Delta N_\theta)+a^2\rho\Delta P_\zeta=\Delta H_5 \tag{7b}$$

$$a(\Delta N_{\theta,\theta}+\rho\Delta N_{\xi\theta,\xi}+2\rho'\Delta N_{\xi\theta})+\omega_\theta\Delta M_{\theta,\theta}+(3\omega_\theta-\omega_\xi)\Delta M_{\xi\theta,\xi}/2$$
$$+[\rho'(3\omega_\theta+\omega_\xi)-\rho\omega_\xi{}']\Delta M_{\xi\theta}/2+a^2\rho\Delta P_\zeta=\Delta H_6 \tag{7c}$$

where

$$\Delta H_4=a\rho\omega_\xi\Delta(\phi_\xi N_\xi+\phi_\theta N_{\xi\theta})+(a/2)\Delta[\phi(N_\xi+N_\theta)]_{,\theta}$$

$$\Delta H_5=a\rho\Delta(\phi_\xi N_\xi+\phi_\theta N_{\xi\theta})_{,\xi}+a\Delta(\phi_\theta N_\theta+\phi_\xi N_{\xi\theta})_{,\theta} \tag{8}$$
$$+a\rho'\Delta(\phi_\xi N_\xi+\phi_\theta N_{\xi\theta})$$

$$\Delta H_6=a\rho\omega_\theta\Delta(\phi_\theta N_\theta+\phi_\xi N_{\xi\theta})-(a\rho/2)\Delta[\phi(N_\xi+N_\theta)]_{,\xi}$$

The effects due to large deformation and large rotations are included in the terms ΔH_1 through ΔH_6.

Under the Kirchhoff-Love hypothesis the strain increment $\Delta\underline{\varepsilon}$ of an arbitrary point of the shell is given by

$$\Delta\underline{\varepsilon}=\Delta\underline{\varepsilon}_m+\zeta\Delta\underline{\kappa} \tag{9}$$

where

$$\Delta\underline{\varepsilon}=\{\Delta\varepsilon_\xi\Delta\varepsilon_\theta\Delta\gamma_{\xi\theta}\},\Delta\underline{\varepsilon}_m=\{\Delta\varepsilon_{\xi m}\Delta\varepsilon_{\theta m}\Delta\gamma_{\xi\theta m}\},\Delta\underline{\kappa}=\{\Delta\kappa_\xi\Delta\kappa_\theta\Delta\kappa_{\xi\theta}\} \tag{10}$$

The underline denotes a vector quantity and the parentheses { } a column vector.

2.2 Constitutive Relations

It is assumed that the constitutive relations of elastoplastic materials can be expressed as

$$\Delta\underline{\sigma}=(\underline{k}^e-\rho^*\underline{k}^p)\Delta\underline{\varepsilon} \tag{11}$$

where $\rho^*=1$ for plastic loading state, otherwise $\rho^*=0$. As has been shown in [6], if we assume Hooke's law for $\underline{k}^e$, we obtain

$$\underline{k}^{p}=\frac{\underline{k}^{e}\underline{S}_{D}\underline{S}_{D}^{T}\underline{k}^{e}}{4E_{p}\sigma_{e}^{2}/9+\underline{S}_{D}^{T}\underline{k}^{e}\underline{S}_{D}} \tag{12}$$

where E_p and σ_e are the plastic tangent modulus and the equivalent stress, respectively. $\underline{S}_D$ denotes the deviator of $\underline{S}=\underline{\sigma}-\underline{\alpha}$ in which $\underline{\sigma}$ and $\underline{\alpha}$ stand for the stress vector and the translation vector of the yield surface. The instantaneous movement $\Delta\underline{\alpha}$ is governed by Ziegler's rule and expressed as follows:

$$\Delta\underline{\alpha}=\frac{3}{2\sigma_{e}^{2}}(1-\chi)\underline{S}\ \underline{S}_{D}^{T}\ \Delta\underline{\sigma} \tag{13}$$

where χ denotes the combination ratio of isotropic and kinematic hardening models.

Applying Eq.(11) to the definitions of the resulting forces $\Delta\underline{N}$ and the resulting moments $\Delta\underline{M}$, we obtain

$$\begin{Bmatrix}\Delta\underline{N}\\ \Delta\underline{M}\end{Bmatrix}=\underline{B}^{e}\begin{Bmatrix}\Delta\underline{\varepsilon}_{m}\\ \Delta\underline{\kappa}\end{Bmatrix}-\begin{Bmatrix}\Delta\underline{N}^{p}\\ \Delta\underline{M}^{p}\end{Bmatrix} \tag{14}$$

where

$$\underline{B}^{e}=\int_{-h/2}^{h/2}\begin{bmatrix}\underline{k}^{e} & \zeta\underline{k}^{e}\\ \zeta\underline{k}^{e} & \zeta^{2}\underline{k}^{e}\end{bmatrix}d\zeta,\quad \begin{Bmatrix}\Delta\underline{N}^{p}\\ \Delta\underline{M}^{p}\end{Bmatrix}=\int_{-h/2}^{h/2}\begin{Bmatrix}\underline{k}^{e}\Delta\underline{\varepsilon}^{p}\\ \zeta\underline{k}^{e}\Delta\underline{\varepsilon}^{p}\end{Bmatrix}d\zeta \tag{15}$$

in which $\Delta\underline{\varepsilon}^{p}$ is the plastic strain increment.

2.3 Fourier Series Expansion

Let us assume that the distributed loads can be expanded into the Fourier series as follows:

$$\Delta\underline{P}=\{\Delta P_{\xi}\,\Delta P_{\zeta}\,\Delta P_{\theta}\}=\sigma_{0}\lambda\sum_{k=0}^{\infty}[\underline{A}_{k}\underline{p}_{k}+\underline{B}_{k}\hat{\underline{p}}_{k}] \tag{16}$$

where $\lambda=h/a$ and σ_0 denotes the reference stress. The coefficients $\underline{A}_k$ and $\underline{B}_k$ are defined by

$$\underline{A}_{k}=\lceil\cos k\theta\ \ \cos k\theta\ \ \sin k\theta\rfloor,\quad \underline{B}_{k}=\lceil\sin k\theta\ \ \sin k\theta\ \ \cos k\theta\rfloor \tag{17}$$

The series of $\underline{A}_k$ are called A-Series and those of $\underline{B}_k$ B-Series. It is assumed further that in accordance with Eq.(16) all the variables can be expanded into the Fourier series. Making use of these Fourier series expansions for the equations obtained above, we arrive at the following governing differential equations for a particular Fourier index k. That is, for the A-Series

$$\underline{R}\ \underline{Z}''+\underline{T}\ \underline{Z}'+\underline{V}\ \underline{Z}=\underline{X} \tag{18}$$

and for the B-Series

$$\underline{\hat{R}}\ \underline{\hat{Z}}'' + \underline{\hat{T}}\ \underline{\hat{Z}}' + \underline{\hat{V}}\ \underline{\hat{Z}} = \underline{\hat{X}} \tag{19}$$

where

$$\underline{Z}=\{u_\xi w u_\theta m_\xi\},\quad \underline{\hat{Z}}=\{\hat{u}_\xi \hat{w} \hat{u}_\theta \hat{m}_\xi\} \tag{20}$$

The coefficient matrices on the left hand sides of Eqs.(18) and (19) are determined by the initial configuration of shell geometry and the elastic constants, while the right hand sides include the external load increments and also the terms resulting from the nonlinear effects due to large deformation as well as plastic deformation.

3. Numerical Method of Solution

An effective solution scheme based on the finite difference approximation was previously proposed[3-6] for the axisymmetric deformation of the shells. Since the governing equations (18) and (19) have the same form as in the axisymmetric case, the solution procedure mentioned above is also applicable to the general case under consideration. Some additional care, however, should be taken to treat the terms involved in the right hand sides of Eqs.(18) and (19). An iterative solution procedure is proposed for this purpose. Namely, for the calculation at each loading step, the values of $\underline{Z}$ and $\underline{\hat{Z}}$ are first assumed and then the computation is iterated until the solutions are converged. The increments thus obtained are integrated numerically by the Euler method. The effectiveness of the proposed method of solution is demonstrated separately in [8,9].

4. Application to Fatigue Life Estimation

The low-cycle fatigue life of the bellows is estimated by using the results obtained from the elastoplastic analysis. Our application is restricted to the case where the bellows is subjected to completely-reversed cycles of axial force. It is assumed that the well known Manson-Coffin equation for the uniaxial low-cycle fatigue:

$$\Delta\varepsilon^p N_f^{\alpha} = C \qquad (\alpha,\ C;\ \text{constants}) \tag{21}$$

where N_f denotes the number of cycles to failure, is valid even for the plane state of stress under consideration, if the plastic strain range $\Delta\varepsilon^p$ is replaced by the equivalent plastic strain range $\Delta\varepsilon^p_{eq}$.

For the bellows problem, $\Delta\varepsilon^p_{eq}$ is defined by

$$\Delta\varepsilon^p_{eq} = (2/\sqrt{3})(\Delta\varepsilon^{p2}_\xi + \Delta\varepsilon^p_\xi \Delta\varepsilon^p_\theta + \Delta\varepsilon^{p2}_\theta)^{1/2} \tag{22}$$

The components of the plastic strain range can be estimated approximately in such a manner as shown in Fig.4, when the axial deflection amplitude w is given. For this estimation, attention is paid to the point at which σ_e obtainable from linear elastic analysis has its maximum value.

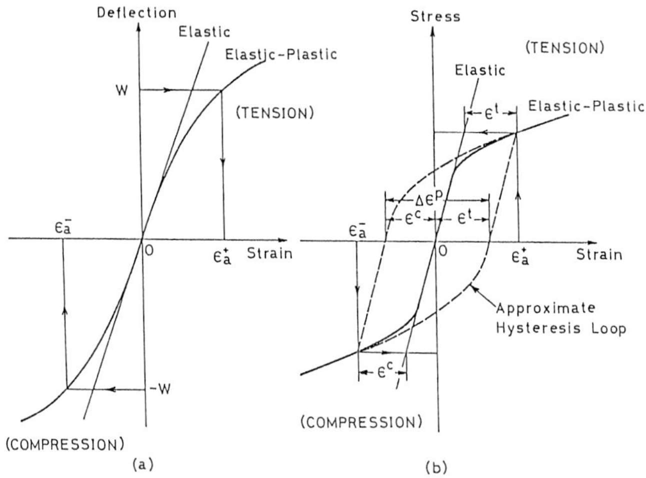

Fig.4 Estimation of plastic range

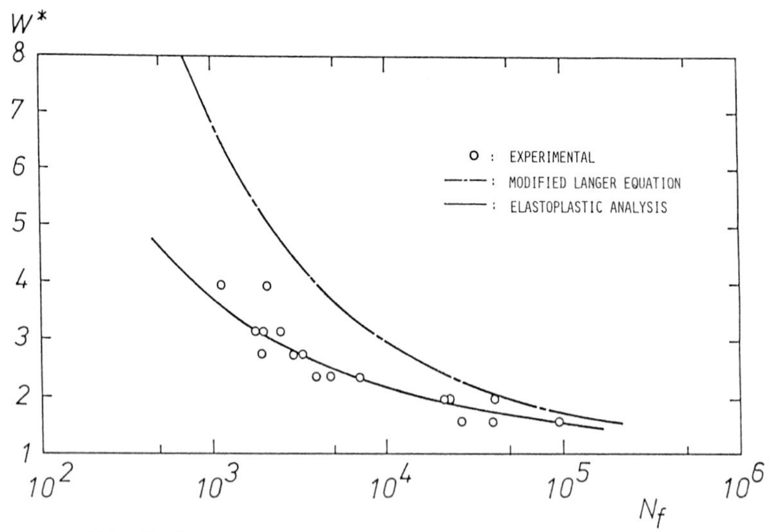

Fig.5 Low-cycle fatigue life of U-shaped bellows

In Fig.5, the low-cycle fatigue lifetimes predicted by elasto-plastic analysis and another method are compared with the experimental ones for the bellows having the following shape parameters:

$$\lambda=h/a=0.0229,\quad \tau=r_u/a=0.216,\quad \delta=b/a=0.928 \tag{23}$$

The notations used are shown in Fig.6. The modified Langer equation in Fig.5 was so determined in [10] that the constants of the Langer equation were calculated as the best fit to some experimental results on the bellows. This equation is expressed by

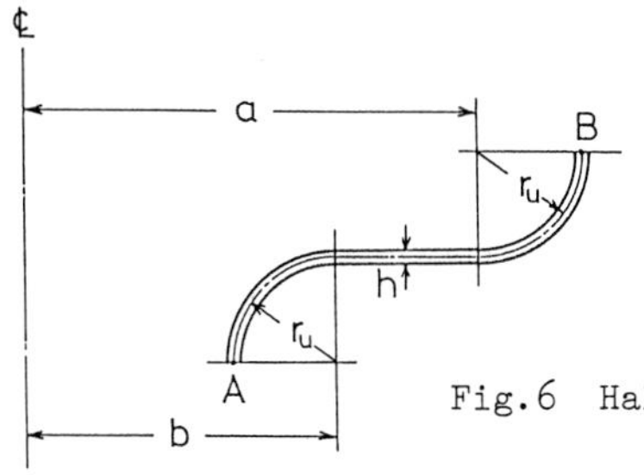

Fig.6 Half pitch of U-shaped bellows

$$N_f=1.68\times10^7/(S-28.1)^2 \tag{24}$$

for which the maximum principal stress obtained from linaer elastic analysis is used in kg/mm^2. The bellows material is assumed to be 18-8 stainless steel for which $\alpha=0.5$, $C=0.65$ and Young's modulus $E=2.1\times10^4$ kg/mm^2. The uniaxial stress-strain property is approximated by a bilinear curve with the following constants:

$$E/\sigma_0=910,\quad E_p/E=0.02 \tag{25}$$

where σ_0 denotes the yield stress. Poisson's ratio is assumed as $\nu=0.3$.

The fatigue life estimation by elastoplastic analysis is carried out for 43 bellows having different shape parameters. The results are summarized in Fig.7 where $\mu=2\tau^2\{\lambda(1+\delta)\}$ and $\alpha_N=N_f/\overline{N}_f$. Here we denote by N_f and $\overline{N}_f$ the fatigue lifetimes predicted by elastoplastic analysis and by the modified Langer equation (24), respectively. The solid curve in Fig.7 is determined as the best fit to the calculated results. This curve is expressed by

$$\begin{aligned}&\alpha_N=0.061+0.314/\mu \quad (0<\mu<1.66)\\&\alpha_N=0.25 \quad (1.66<\mu)\end{aligned} \tag{26}$$

Comparison is made in Fig.8 between Eq.(26) and the experimental results. The solid curve indicates Eq.(26) and the mark o the experimental results.

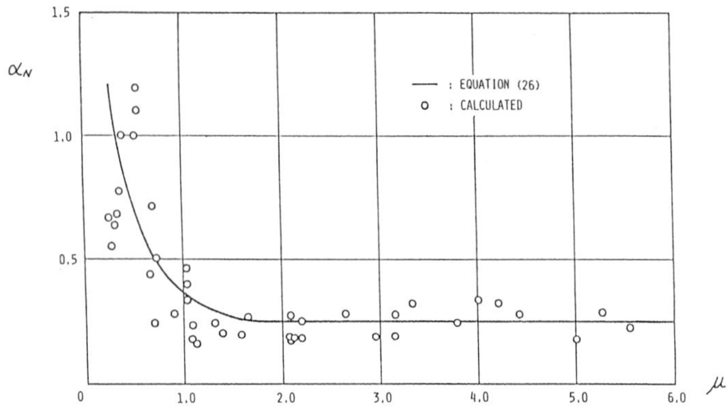

Fig.7 Fatigue lifetime ratio

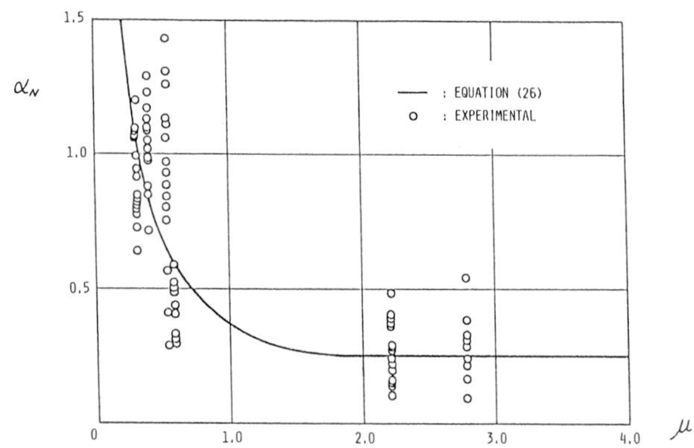

Fig.8 Comparison of Eq.(26) with experiment

5. Conclusions

We can summarize our design scheme for the low-cycle fatigue of the bellows as follows: If the maximum elastic stress S is obtained from the design charts or so, the lifetime $\overline{N}_f$ can be calculated by use of Eq.(24). On the other hand, the lifetime ratio α_N can be obtained using Eq.(26). Then, we can find the actual lifetime N_f as $N_f=\alpha_N\overline{N}_f$.

References

[1] Stricklin, J.A., et al.; Trans ASME, Ser.J,96-2(1974),87.
[2] Zienkiewicz, O.C.; The Finite Element Method in Eng. Sci.,(1971).
[3] Hamada, M. and Tanaka, M.; Bull. JSME, 14-74(1971),724.
[4] Hamada, M. and Tanaka, M.; Bull. JSME, 15-87(1972),1029.
[5] Hamada, M. and Tanaka, M.; Bull. JSME, 17-103(1974),41.
[6] Tanaka, M.; Ing.-Arch., 41-5(1972),342.
[7] Sanders, J.L.,Jr.; Quart. Appl. Math., 21(1963),21.
[8] Hamada, M. and Tanaka, M.; Bull. JSME, 23-177(1980),345.
[9] Hamada, M. and Tanaka, M.; under submission.
[10] Hamada, M. and Takezono, S.; Bull. JSME, 10-40(1967),611.

Neutral and Shakedown-Type Domains for Axisymmetric Plates and Shells under Cycles of Pressure and Temperature

Zenon WASZCZYSZYN and Ewa PABISEK

Institute of Structural Mechanics
and Computer Center
Technical University of Cracow
31-155 Kraków, ul.Warszawska 24, POLAND

Summary

Incremental equations for elastic-plastic material with combined isotropic-kinematic strain-hardening and equations of large deflections of shells of revolution are used. The shooting method is applied to integrate the equations along the Load Control Program /LCP/ path. Neutral Domain Curves/NDC/ are evaluated for prescribed points of LCP. by means of an approximate method of superposition of an elastic-plastic solution and elastic unloadings. Skakedown-type Domains /SD/ are calculated for the radial LCP paths.

1. Introduction

The analysis of structures subjected to multiparametric loads is one of the most important design problems. In the case of elastic-plastic material the shakedown theorems can be applied for evaluating approproate limit surfaces in the load-parameter space [1]. Besides serious theoretical difficulties /finding of appropriate fields of residual stresses/ application of these theorems is drastically limited by considerable simplifications of the mathematical model /ideally elastic-plastic material, small displacements/.

The mentioned difficulties can be overcome if a relation between the load parameters is known. Exploration conditions of many structures enable us to formulate the so-called Load Control Program /LCP/. In the case considered the LCP can be written in the implicit form:

$$F(P,T) = 0 \in C^{0}, \qquad /1.1/$$

where P, T are pressure and temperature parameters respectively.

For every point M_i on the LCP path the Neutral Domain /ND/ of elastic unloadings can be evaluated - fig.1a. This domain is bounded by the Neutral Domain Curve /NDC/ of the equation

$$\Phi_i(P,T) = 0 . \qquad /1.2/$$

The shape of the NDC depends both on the development of yielding zones of the structures and on all the previous history of the loading process along the LCP path.

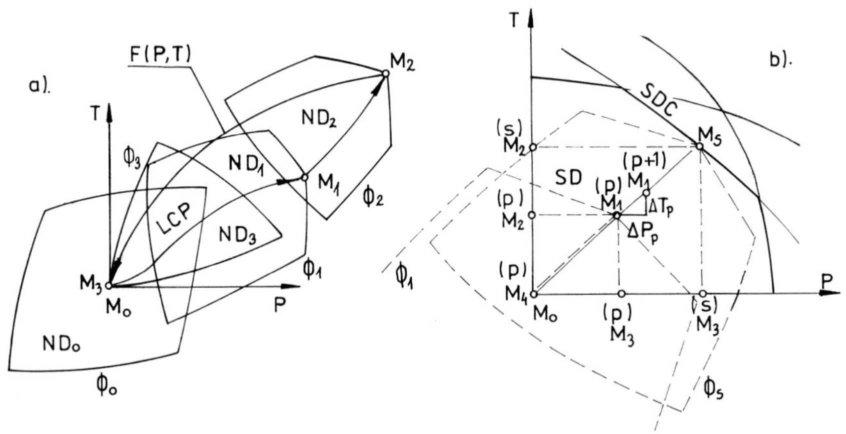

Fig.1. Load Control Program /LCP/, Neutral Domains /ND/ and Shakedown-type Domain /SD/.

Besides NDC the Shakedown-type Domain /SD/ will be evaluated for a family of LCP paths. The idea is to find the points $M_s = M(P_s, T_s)$ of the SD curve for which no increments of plastic strains are produced if the unloading paths are in the rectangle $\{0 \le P \le P_s,\ 0 \le T \le T_s\}$. In fig.1b an example of the SD curve is shown for the radial NDC paths.

2. Assumptions

1. The provlem is circularly symmetric so only one independent geometrical variable is used.

2. The theory of thin shells /the Kirchhoff hypotheses/ and moderately large deflections are assumed.

3. The ideal sandwich, symmetric cross-section is of constant thickness and constant distance between the faces.

4. Material of the faces is isotropic, elastic-plastic, compressible with combined kinematic-isotropic strain-hardening. The yield point and the coefficients of strain-hardening are the known functions of temperature.

5. The problem is quasistatic with a known distribution of temperature. Changes of pressure and temperature are defined by two load parameters P and t .

3. Basic equations

The set of large deflection equations of thin shells of revolution can be written in the quasi-linear form [2]:

$$\begin{aligned} &\underline{y}' = \underline{f}(\underline{y}, \vartheta_1, k_1, n_2, m_2; \xi, P), \quad \underline{y}, \underline{f} \in R^6 \\ &\underline{\dot{Q}} = \underline{D}\, \underline{\dot{q}} - \underline{G}\dot{\vartheta}, \quad \underline{Q}, \underline{q} \in R^4, \end{aligned} \qquad /3.1/$$

where the state vector $\underline{y}$ /cf.fig.2a/ and the vectors of generalized stresses $\underline{Q}$ and generalized strains $\underline{q}$ are introduced:

$$\begin{aligned} &\underline{y} = \{u, v, \varphi, n_1, t, m_1\}, \\ &\underline{Q} = \{n_1, n_2, m_1, m_2\}, \quad \underline{q} = \{\vartheta_1, \vartheta_2, k_1, k_2\}. \end{aligned} \qquad /3.2/$$

Eqs /3.1/ are written in dimensionless quantities. The full set of ordinary differential equations is separated with respect to the space and time derivatives, $(\,)' = \partial/\partial\xi$ and $\dot{(\,)} = \partial/\partial\tau$ respectively. Such a separation is possible since the material coordinate $\xi \in [\xi_0, \xi_L]$ and the quasi-time τ are used as independant variables.

Eqs /3.1/$_1$ formulate a two-point BV problem of the range n=3, i.e. three boundary conditions are to be added at the edge points ξ_0 and ξ_L of the shell. The boundary conditions have to be related to components of the state vector $\underline{y}$ /fig.2a/.

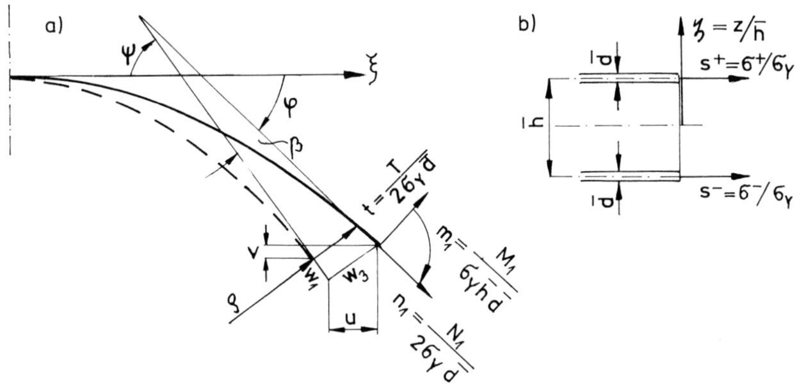

Fig.2. Components of the state vector and ideal sandwich cross-section of a shell.

The BV problem can be reduced to n=2 if no restrictions against the vertical deflections in the points ξ_0 either ξ_L of the plate are applied [3].

Owing to the ideal sandwich cross-section the components of the vector $\underline{Q}$ can be expressed by stresses in the upper and lower faces /fig.2b/:

$$n_j = (s_j^+ + s_j^-)/2 \ , \quad m_j = (s_j^+ - s_j^-)/2 \ , \qquad /3.3/$$

where j=1,2 denotes the principal directions which coincide with the lines of curvatures, meridional and circumferential respectively.

The Kirchhoff-Love hypothesis of a normal segment enables writing down the formulae for components of the vector of generalized strains $\underline{q}$:

$$\ni_j = (e_j^+ + e_j^-)/2 , \quad k_j = e_j^+ - e_j^- . \qquad /3.4/$$

The components of matrices $\underline{D}$ and $\underline{G}$ in the constitutive relation $/3.1/_2$ can be expressed as a linear combination of the components of E_{jk}^r and F_j^r which are used in the local physical relation [3]:

$$\dot{s}_j^r = (E_{jk}\, \dot{e}_k - F_j \dot{\vartheta})^r \quad \text{for} \quad j,k=1,2;\ r=+,- . \qquad /3.5/$$

Besides the increments of stresses $\dot{s}_j^r$ and strains e_j^r the increment of temperature is used, according to an assumed field of temperature:

$$\dot{\vartheta} = \dot{T}(\Gamma)\, f(\xi) \quad . \qquad /3.6/$$

The matrices E_{jk}^r and F_j^r depend both on the relative stresses $s_j^{\circ r} = s_j^r - a_j^r$ and on the development of the yielding measured by the Odqvist parameter $\bar{e}^r$. Isotropic and kinematic strain-hardening are taken into account through some functions $k(\bar{e},\vartheta)$ and $C(\bar{e},\vartheta)$, which appear in the Huber-Mises yield condition and in the Ziegler law:

$$s_e^{\,2} - 3\, k^2(\bar{e},\vartheta) = 0 , \quad s_e^{\,2} = s_1^{\circ 2} - s_1^{\circ}\, s_2^{\circ} + s_2^{\circ 2} ,$$
$$\dot{a}_j = C(\bar{e},\vartheta)(2\, \dot{e}_j^p + \dot{e}_k^p) , \quad j \neq k , \qquad /3.7/$$

where a_j describes the position of the yield curve centre and $\dot{e}_j^p$ are increments of the plastic part of the strain increment $\dot{e}_j$.

The stresses and strains are related to appropriate values at the yield point for a unidimentional tension, i.e. $s_j = \sigma_j/\sigma_Y$, $e_j = \varepsilon_j/\varepsilon_Y$.

In [3] appropriate formulae have been derived for the components of the matrices E_{jk}, F_j which depend on the type of the loading process. In the case when the increment of the Odqvist parameter $\Delta \bar{e}_m = \bar{e}(\tau_{m+1}) - \bar{e}(\tau_m) \geq 0$ the process is active and equations of the plastic flow are applied. If $\Delta \bar{e}_m = 0$ and $s_e^2 < 3k^2$ then local unloading occurs and linear equations of the theory of elasticity are used.

4. Methods of solution

The integration of Eqs /3.1/ along the LCP path is carried out by an incremental procedure with respect to the quasi-time τ. This parameter can be identified as a parameter of the curve /1.1/, i.e. the load parameters can be expressed explicitly:

$$P = f_1(\tau), \quad T = f_2(\tau) . \qquad /4.1/$$

A modified shooting method [4] is applied to solve the BV problem for the fixed value $\tau = \tau_m$.

After the point M_i on the LCP is reached the full set of information is stored. That enables us to evaluate the ND curve applying procedures described in detail in [3,5].

An exact method was worked out in [3]. The method depends on the integration of the full set of equations /3.1/ for finding iteratively subsequent points of the NDC.

Such an exact method turned out to be very time consuming so an approximate approach was proposed in [5]. The main idea of the approximate method lies in the computation of basic elastic solutions taking into account yielding of the structure. Such solutions are superposed with the elastic--plastic solution reffered to the point M_i of the LCP.

The SD curve is computed integrating Eqs /3.1/ along the radial path $M_0 \; M_1 \; P_1, T_1$. Then unloading is performed for paths pointed out in fig.1b: $\overset{p}{M_1} \rightarrow \overset{(p)}{M_2}(0, T_1)$, $\overset{(p)}{M_1} \rightarrow \overset{(p)}{M_3}(P_1, 0)$ $\overset{(p)}{M_1} \rightarrow M_4 \; (0,0)$. If at the points $\overset{(p)}{M_i}$ for $i=2,3,4$ the yield curve is not reached in any part of the structure / no secondary yielding occurs/ the BVP is solved for the subsequent point $\overset{(p+1)}{M_1} = M_1\,(P_1 + \Delta P_p,\ T_1 + \Delta T_p)$ and the external unloading process is repeated. After the yield curve is crossed for any $\overset{(p+1)}{M_1}$ the point M_s on the SD curve is found iteratively. In fig.1b this point is related to the unloading down to point $\overset{(s)}{M_2}$.

5. Numerical examples

Only some results are given heredown to illustrate the proposed approaches. The more complicated examples were presented on the Symposium.

A good agreement has been found between the exact and approximate method so in fig.3 only the approximate ND curves are shown. Only kinematic linear strain-hardening of the

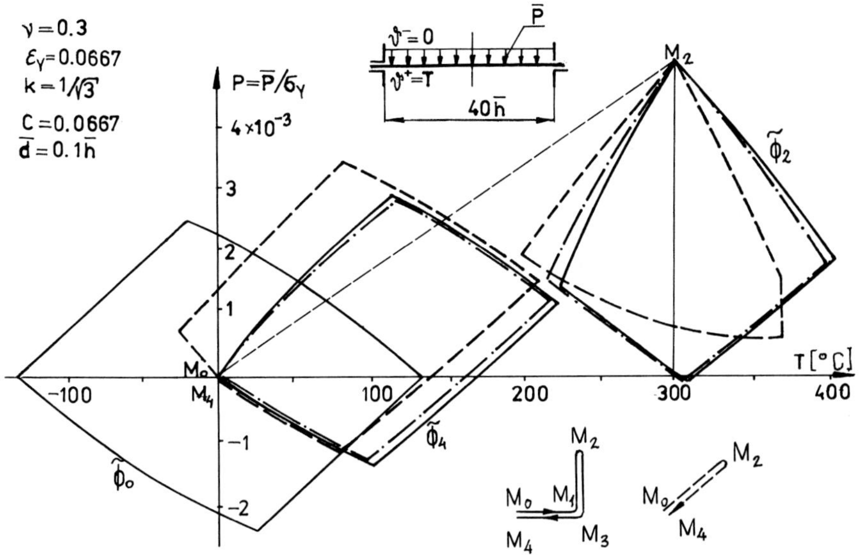

Fig.3. Neutral Domain Curves for Load Control Programs: —— single cycle A ,—·— triple cycle A , ––– single cycle B.

coefficient C=0.0667 is considered. The ND curves have been computed for two kinds of single and a triple cycles.

Boundary conditions and dependance of the yield point on temperature influance significally the NDC. The ND curves can be concave, this being caused by large deflections.

The SD curves are shown in fig.4 for plates with different boundary conditions with respect to the membrane forces. The shakedown-type domains are bounded by the SDi curves which are related to the points M_i^s on the unloding paths.

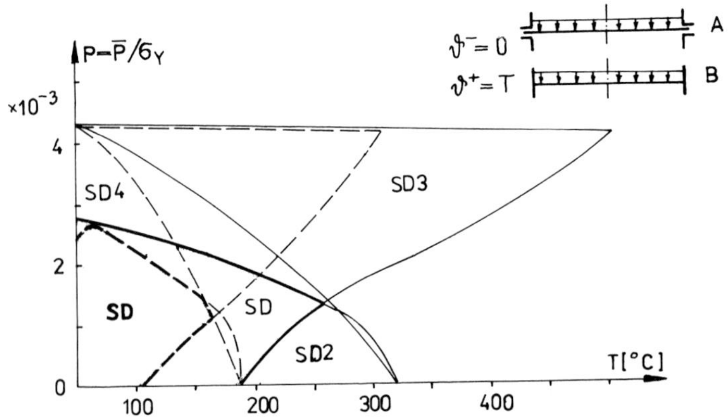

Fig.4. Shakedown-type Domains for plates: —— A , – – – B and other data as in fig.3.

Literature

[1] Życzkowski, M.: Combined loads in the theory of plasticity /in Polish, to be translated in English/, Warsaw: PWN 1973.

[2] Waszczyszyn, Z.: Arch. of Mechanics 24 /1972/ 483-498.

[3] Waszczyszyn, Z.; Pabisek, E.: Nucl. Eng./Design 45 /1978/ 173-183.

[4] Waszczyszyn, Z.: Symp. franco-polon., Nice 1974, Warsaw: PWN 1977, 465-477.

[5] Waszczyszyn, Z.; Pabisek, E.: SMIRT 5, Berlin 1979, Rep. L11/7.

Accuracy of Numerical Solutions of Problems Involving Physical Non-Linearities

Y. YAMADA

Institute of Industrial Science

University of Tokyo, JAPAN

Summary

Meeting the demand of material scientists, finite element specialists should strive to improve the accuracy of inelastic analysis without arousing an immoderate increase of total computer time and cost. From the efforts directed to the target and the recent progresses thus achieved, some typical examples are introduced. Notices are also given to the usefulness and cost effectiveness of small computer system for developing basic routines for the solution of nonlinear problems.

SOLUTION OF BENCHMARK PROBLEMS

Activities in Japan of verification and/or qualification of inelastic analysis programs in 1973-78 were reported by the author [1, 2]. The cooperative work is succeeded to JSME RC-Subcommittee 55 on Design Application Procedure of Inelastic Analysis and the efforts are focused on the participation in the international benchmark analysis programs. The major subjects the Subcommittee concerns at present time are the problems proposed by ORNL [3], PVRC [4] and for second International Conference on Numerical Methods in Fracture Mechanics held in Swansea on July 6-11, 1980.

Fig.1 is a summary of the solutions obtained by the JSME Subcommittee for benchmark problem 1 of Swansea Conference [5] and shows the crack opening displacements in compact tension specimen under the practically plane stress condition. Although considerable plasticity occurs in this problem, the agreement of results from five organizations using different programs is satisfactory. It is concluded that we reached to the stage whereon good correlations can be expected between solutions of elastic-plastic deformations, provided the problem is posed properly.

DEPENDENCE OF SOLUTION ON COSTITUTIVE MODELING

It is evident even for the case of elastic-plastic analysis numerical results show considerable deviation or scatter when the basic assumptions on material behavior or constitutive modeling differ. In order to correlate the results due to various postulates, the JSME Subcommittee set up a benchmark problem of Fig.2. Two combined tension-torsion stress paths having sharp bends are specified and the comparison of solutions for strain paths is requested. Analytical solutions are worked out for isotropic and kinematic hardening materials and utilized as the basis of verification of numerical analysis routines. The scheme is being extended to the combined hardening material and the fraction or sublayer model. It is considered that further sophistication of modeling would not be in place before it becomes clear that all the elementary postulates fail in the prediction of the actual experimental results. As can be seen from the strain path of Fig.3 corresponding to the stress path ABCD in Fig.2, there is a room to improve or adjust the prediction by modifying the hardening rules as well as the approximations adopted in the description of material behavior.

In contrast to the case of elastic-plastic analysis, solutions show significant discrepancies as the effect of creep becomes important. To draw attention to these facts the JSME Subcommittee established a couple of benchmark problems which are classified into monotonic proportional loading, proportional stress reversal, and nonproportional loading with altenation or rotation of stress directions (see Fig.4).

As regards the time dependent mechanical behavior, it is very often that the creep deformation is overestimated compared to the experimental results. It is considered that this is due to the underestimate of the stress relaxation. For example, if the Norton's creep law $\dot{\varepsilon} = \dot{\varepsilon}_0(\sigma/\sigma_0)^n$ is adopted, the stress state is strongly affected by the value n which is determined by the creep test but may assume a smaller value for the stress relaxation.

CONVERGENCE OF SOLUTIONS

Incorporation of appropriate iterative procedure is crucial in nonlinear analysis for the purpose of improving the accuracy of computation. In this respect, recently proposed quasi-Newton method [6] and accelerated modified Newton-Raphson iteration [7] deserve considerations. From the practical view point, determination of incremental step size is important as well in the elastic-plastic analysis when sharp bend manifests in the stress path. The stress state at the notch root and the crack tip frequently exhibits such discontinuous changes as the plastic enclave or crack grows. Yokouchi and Ishii [8] proposed a method to restrict the size of load increment by referring to the error of the estimated equivalent strain increment or the equivalent stress increment for the von Mises material.

A MICROCOMPUTER SYSTEM AND NONLINEAR ANALYSIS

A microcomputer system is built and being tested for its usefulness for the solution of nonlinear problems [9]. FEMN is the program being developed in the parallel efforts. The restart capability and the program organization of FEMN make the system versatile, as the user can define willingly the sequence of loading as well as the nonlinear material properties concerned. It is hoped that wide dissemination of the finite element method through small computer system contributes to the deeper understanding of implications of physical non-linearities in structural analysis.

REFERENCES

1 Yamada, Y.: Research and development of inelastic analysis procedures for reactor component design in Japan. Nuclear Engng. and Design, 51 (1978) 85-96.

2 Yamada, Y.: Subsequent activities in Japan of inelastic structural analysis for elevated temperature design. 2nd Int. Seminar on Inelastic Analysis and Life Prediction in High Temperature Environment, 5th SMiRT Conf., Berlin (1979).

3 Corum, J.M.; Wright, W.B. (eds.): Pressure vessel and piping: Verification and qualification of inelastic analysis computer programs. ASME (1975).

4 Benchmark Problem No.3 of PVRC Subcommittee on Elevated-Temperature Design: Nozzle-to-sphere attachment region undergoing

thermal transient, axial and pressure loadings (1978).

5 Compact tension specimen analysis. Report from RC-SC 55 Sub-committee on Design Application Procedure of Inelastic Analysis, JSME. 2nd Int. Conf. on Num. Meth. in Frac. Mech., Swnasea (1980).

6 Matthies, H.; Strang, G.: The solution of nonlinear finite element equations. Int. J. num. Meth. Engng., 14 (1979) 1613-1626.

7 Crisfield, M.A.: A faster Newton-Raphson Iteration. Comp. Meth. Appl. Mech. Engng., 20 (1979) 267-278.

8 Yokouchi, Y.; Ishii, A.: Error estimate of tangent stiffness method and a control of step size of load increment. paper for JSME meeting, Gifu, Japan (1980), in Japanese.

9 Yamada, Y.: Microcomputer system for medium-sized and experimental finite element analysis. Paper presented at Symp. on Computational Mech. in Nonlinear Struct. and Solid Mech., Washington, D.C., USA (1980).

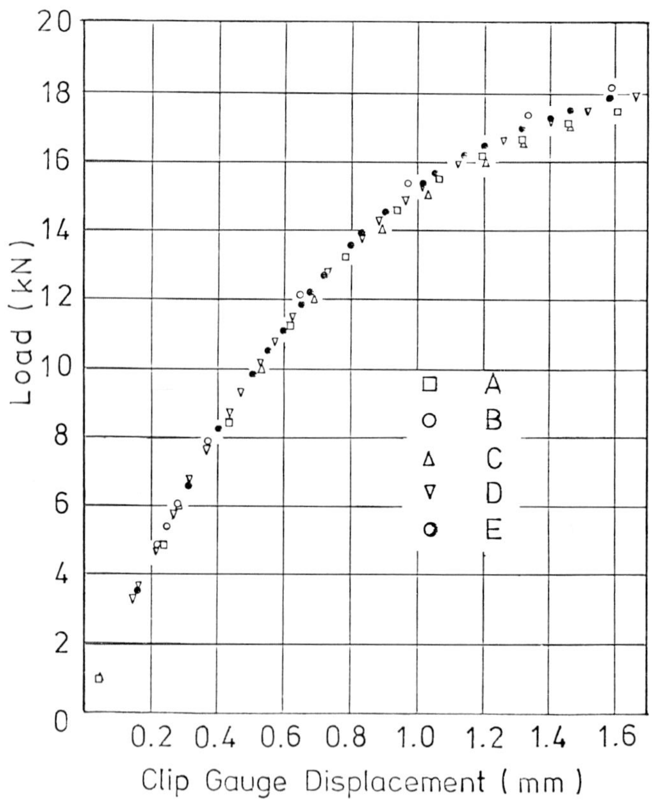

Fig.1 Load versus clip gauge displacement of compact tension specimen

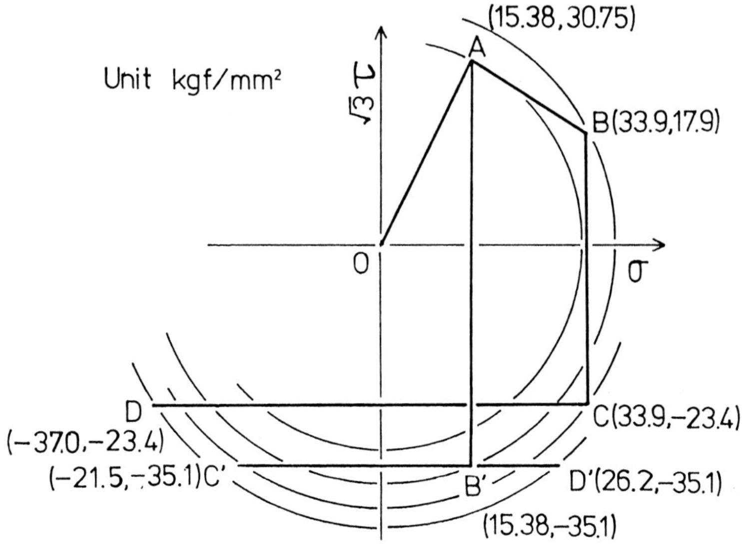

Fig. 2 Combined tension - torsion stress paths

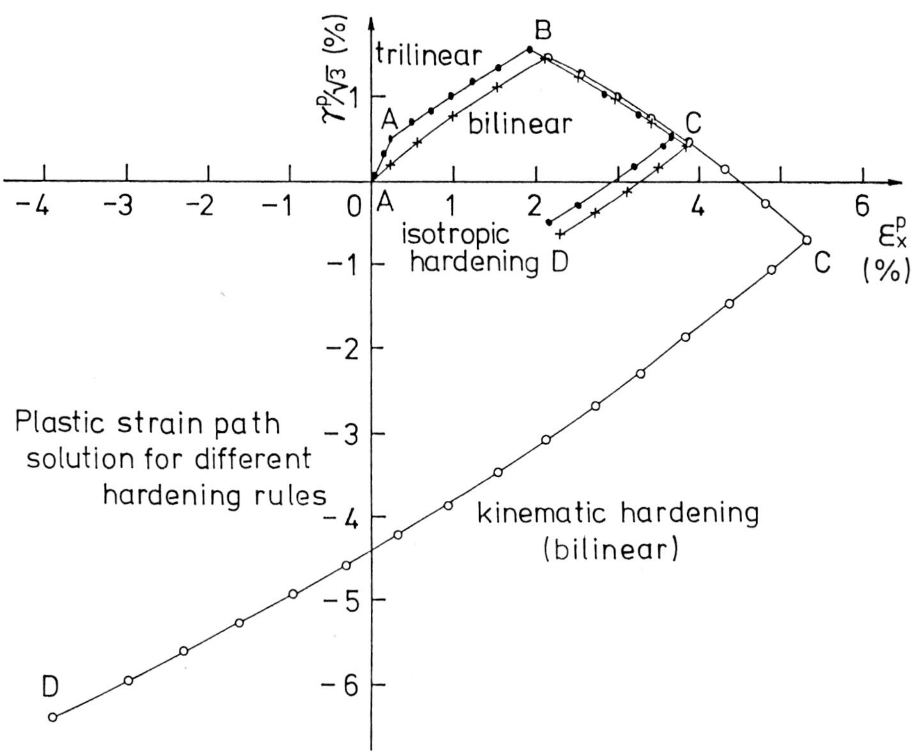

Fig. 3 Plastic strain path solution for different hardening rules

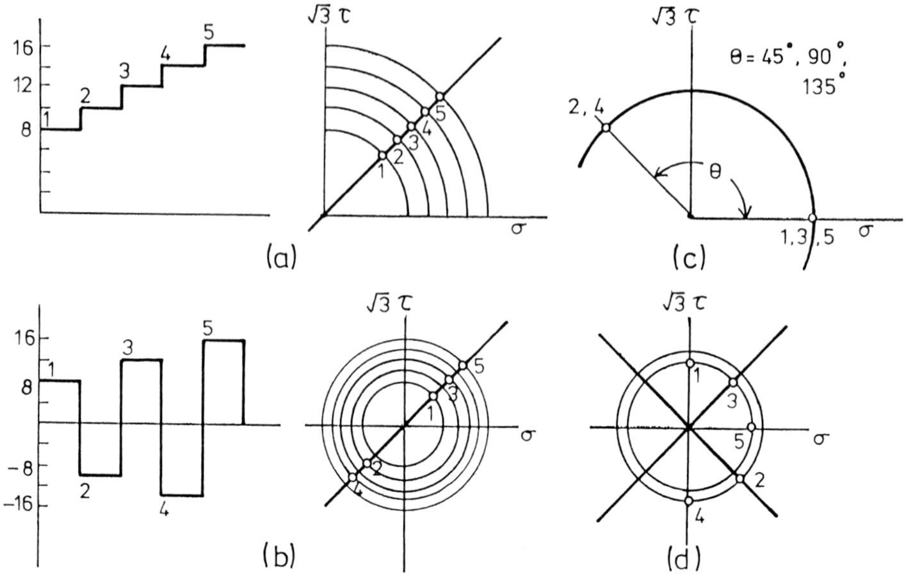

Fig.4 Creep test under various stress conditions(a) – (d)

Physical and Phenomenological Model with Non-Linearity in Ductile Fracture and Fatigue Crack Growth

T. YOKOBORI* and A.T. YOKOBORI, JR.**

* Research Institute for Strength and Fracture of Materials, Tohoku University, Sendai, JAPAN

** Department of Mechanical Engineering II, Tohoku University, Sendai, JAPAN

Summary

This paper concern the ductile fracture with respect to large scale non-linear plasticity, and concern the fatigue crack growth at not high temperatures with respect to small scale non-linear plasticity.

Concerning the ductile fracture, a physico-phenomenological model has been presented both for slipping -off type and for separation type, respectively. For separation type fracture, the piling-up of dislocations problems have been studied by computer simulation based on dislocation group dynamics.

Concerning the fatigue crack growth, a physico-phenomenological model has been presented. The crack growth rate formula based on this model is in common with any of physical models, such as, nucleation model, dislocation group dynamics model or vacancy condensation model as far as stress intensity dependence is concerned.

Introduction

For unstable fracture among the ductile fracture accompanying large scale non-linear plasticity a criterion has been proposed based on energy balance condition.[1)2)] However, when non-linearity becomes more larger, the contribution from elastic strain energy becomes smaller and thus the fracture criterion based on energy balance condition may be difficult to be derived. Thus, for the ductile fracture with much larger scale non-linear plasticity, as phenomenological model the following models of the two types may be assumed.: slipping-off model and separation model. In the following two sections, physico-phenomenological model were presented for the fracture of these two types, respectively.

On the other hand, in fatigue crack growth at not high temperatures the crack tip may accompany small scale non-linear plasticity. Nevertheless, fatigue crack growth rate is not characterized by linear fracture mechanics, but is described by small scale non-linear plastic stress distribution at the crack tip.[3)-6)] For such case, several physical models have been proposed. In the last section of this article, a physico-phenomenological model has been presented in common with any of physical models, such as, nucleation model,[3)4)] dislocation group dynamics model[5)] or vacancy condensation model[6)] as far as stress intensity dependence of fatigue crack growth rate is concerned.

Ductile Fracture of Slipping-off Type

For usual COD model the physical meaning has not been made clear. For instance, COD model may be interpreted phenomenologicaly the same[7)] essentialy as the Ludwik's concept[8)] of fracture of notched specimen except that the latter assumes the plastic strain averaged over the notch part section at fracture instead of the strain at the notch root. However, the hypotheses is not physicaly clear that we assume the definite relation between completely brittle fracture strength and pre-plastic strain. On the other hand, Cottrell's physical interpretation[9)] may be that fracture will occur by slipping-off of the material, that is, columnar part, between the crack tip and the void nucleated. (Fig.1). However, the formulation based on the physical basis does not seem suitable for computing and comparison with experimental data.

Based on computer simulation on dislocation group dynamics with dislocation emission[10)-13)] from the crack tip, COD is given by[5)]

$$COD \simeq 2\alpha nb , \tag{1}$$

where b = Burgers vector, and n = the number of dislocations emitted from the crack tip until some certain time and $\alpha(<1)$ is orientation factor. If we assume that the slipping-off of the columnar part (Fig.1) will occur when the number of dislocations emitted from the crack tip exceeds some critical number n_c; then eqn.(1) may be a criterion for the ductile fracture of this type.

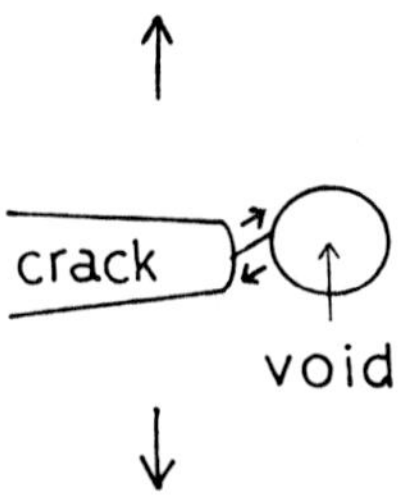

Fig.1. Ductile fracture of slipping-off type.

Ductile Fracture of Separation Type

(1) Why microscopic concept is necessary?

In this section let us consider the ductile fracture caused by the separation of the materials with large scale yielding region or of the columnar type materials between the crack and the near-by void tip accompanying with large scale non-linear plasticity. In this case, whatever the microscopic model may be, the separation should occur by local tensile stress, but not local shear stress. This point is quite different from the model for ductile fracture of slipping-off type. That is, the requisite for fracture is that the local tensile stress at the crack tip exceeds the atomic bonding force.

On the other hand, even under the condition of large scale yielding, the local macroscopic tensile stress, σ_ℓ can not amount to at most ten times of macroscopic yield stress, σ_Y. For instance, under plain strain for perfect plastic materials, σ_ℓ/σ_Y can not exceed about three,[14] and for strain hardening materials, σ_ℓ/σ_Y assumes at most[15] about 10 as shown in Table 1.† Therefore, by macroscopic mechanical treatment alone, high local stress cannot be obtained enough to breaking the atomic bonding force at or near the crack tip. In turn, the concept of some micro-

Table 1. The maximum value of σ_ℓ/σ_Y for strain hardening materials.(calculated by eqn.(15) in Ref.15)

crack length, mm	crack tip radius, mm	strain hardening exponent	σ_ℓ/σ_Y
3	0.01	0.35	4.36
3	0.001	0.35	7.93

scopicaly local stress concentration should necessarily be involved. Now, whatever the micromechanism may be, the stress concentration of this type may be given phenomenologicaly by dislocation pile-up. On the other hand, most of analyses on dislocation pile-up in literatures concern the equilibrium pile-up of dislocations.[16] Therefore, in this section, we calculate the microscopicaly local tensile stress in terms of dislocation group dynamics with its emission from the source.

(2) Numerical analysis on the dynamic piling-up of dislocations emitted from the source.

There are some papers[17)18)] on dislocation pile-up, but not so ready for the application to fracture mechanics. Thus, in this section numerical analysis was carried out on the dynamic piling-up of dislocations emitted from stressed source and on the dynamic stress intensity factor caused by crystal lattice dislocations pile-up.

(2)-1 Model, basic equation and analysis

Until the lead dislocation arrives at a barrier, such as, grain boundary, dislocations will emit from a stressed source and move ahead freely but with interaction between dislocations as shown in Fig.2. After the arrival of the lead dislocation at the barrier, it is locked and the trailing dislocations pile up against the barrier as shown in Fig.3.

The power relation between stress τ and velocity v was assumed for each individual dislocation in a linear array as follows.

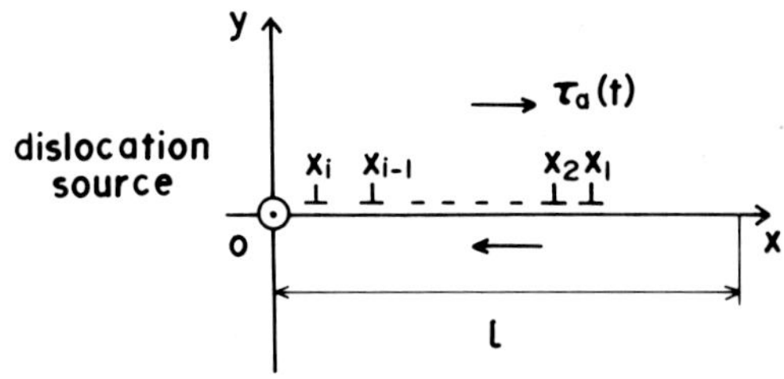

Fig.2. Dislocations moving ahead emitted from the source.

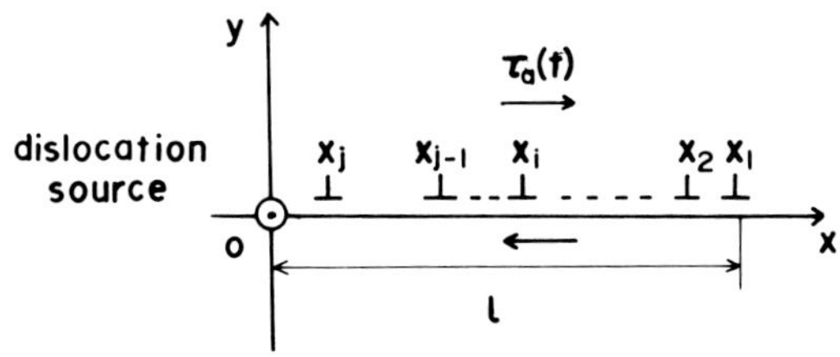

Fig.3. Dynamicaly piling-up of dislocations emitted from the source.

$$v_i = \frac{dx_i}{dt} = M\tau_{eff,i} \, , \qquad (i = 1, \text{-----} n) \qquad (2)$$

where $\tau_{eff,i}$ is the effective stress on individual dislocations in terms of shear stress, m is a material constant and $M = v_0 \, (1/\tau_0^*)^m$. In the equation, τ_0^* is a material constant representing the stress required to give a dislocation velocity $v_0 = 1 \text{ cms}^{-1}$ and i is the index number of the dislocation in order of emitting from the source. Denoting by ℓ the distance between dislocation source and the barrier, then $\tau_{eff,i}$ is written as eqn.(3) for $x_1 < \ell$ and as eqn.(4) after the arrival of the lead dislocation at the barrier respectively:

$$\tau_{eff,i} = \tau_a(t) + A \sum_{\substack{j=1 \\ i \neq j}}^{n} \frac{1}{x_i - x_j} \, , \qquad (x_1 < \ell, i=1,2,3,\text{---}n) \qquad (3)$$

and

$$\tau_{eff,i} = \tau_a(t) + A\left(\frac{1}{x_i - \ell} + \sum_{\substack{j=2 \\ i\neq j}}^{n} \frac{1}{x_i - x_j}\right), \quad (x_1=\ell, i=1,2,3,\text{--}n) \tag{4}$$

where A is Gb/2π(1−ν) for edge dislocation and Gb/2π for screw dislocation. G=shear modulus, say, 8100kg/mm^2, ν = poisson's ratio, say, 0.3, b = burgers vector, say, 3×10^{-7}mm, $\tau_a(t)$ = applied stress in terms of shear stress. In this section we assumed the case for the constant applied stress rate, $\dot{\tau}$, that is, $\tau_a(t) = \dot{\tau}t$. $\dot{\tau}$ is assumed, say, as $10^3\text{kg/mm}^2\text{s}$. Initial condition is that at t=0, n=1 and x_1=0, that is, at t=0 one dislocation is at the source. On the other hand, the effective stress exerted on a source, $\tau_{eff,s}$ is written as eqn.(5) for $x_1<\ell$ and as eqn. (6) after the arrival of the lead dislocation at the barrier respectively:

$$\tau_{eff,s} = \dot{\tau}t - A\sum_{j=1}^{n}\frac{1}{x_j}, \quad (x_1<\ell) \tag{5}$$

$$\tau_{eff,s} = \dot{\tau}t - A\left(\frac{1}{\ell} + \sum_{j=2}^{n}\frac{1}{x_j}\right). \quad (x_1=\ell) \tag{6}$$

In calculation, it is convenient to rewrite eqns.(1)~(6) in terms of non-dimensional form: thus let us take

$\dot{\tau}_0$ = the specified rate of stress application, say, $1\text{kg/mm}^2\cdot$sec,

ℓ = the distance between dislocation source and the barrier, say, 10^{-2}mm,

t_0 = the time required for an isolated dislocation to run by the distance ℓ, under $\dot{\tau}_0$.

Then we get

$$t_0 = [(m+1)\ell / M\dot{\tau}_0^m]^{1/(m+1)} \tag{7}$$

By using the following non-dimensional variables

$$\theta = t/t_0 , \tag{8}$$

and

$$s_i = x_i/\ell \tag{9}$$

and eqn.(7), then eqn.(2) reduces to

$$\frac{ds_i}{d\theta} = (m+1)\left(\frac{\dot\tau}{\dot\tau_0}\theta + \frac{A}{\dot\tau_0 t_0 \ell}\sum_{\substack{i=1\\ j=i}}^{n}\frac{1}{s_i - s_j}\right)^m , \quad (s_1 < 1) \tag{10}$$

and

$$\frac{ds_i}{d\theta} = (m+1)\left\{\frac{\dot\tau}{\dot\tau_0}\theta + \frac{A}{\dot\tau_0 t_0 \ell}\left(\frac{1}{s_i - 1} + \sum_{\substack{j=2\\ j=i}}^{n}\frac{1}{s_i - s_j}\right)\right\}^m . \quad (s_1 = 1, i = 2, -- n) \tag{11}$$

Also eqns.(5) and (6) reduce to, respectively,

$$\frac{\tau_{eff,s}}{\dot\tau_0 t_0} = \frac{\dot\tau}{\dot\tau_0}\theta - \frac{A}{\dot\tau_0 t_0 \ell}\sum_{j=1}^{n}\frac{1}{s_j} , \quad (s_1 < 1) \tag{12}$$

and

$$\frac{\tau_{eff,s}}{\dot\tau_0 t_0} = \frac{\dot\tau}{\dot\tau_0} - \frac{A}{\dot\tau_0 t_0 \ell}\left(1 + \sum_{j=2}^{n}\frac{1}{s_j}\right) . \quad (s_1 = 1) \tag{13}$$

The method of computation is the similar as used in previous paper.[10)-13)] That is, selecting a time increment $\delta\theta$, δs_i of the increment of s_i of each of the dislocations in the array was calculated by the Runge-Kutta-Merson method. When $\tau_{eff,s}$ becomes equal to the source activation stress τ_s which is assumed zero in this section, then a new mobile dislocation was introduced at the dislocation source. Thus the number of dislocations increases and, therefore, the number of simultaneous differential equations of eqn.(10) or eqn.(11) increases by one corre-

spondingly, and the procedure was iterated. Eqn. (10) and eqn. (11) were programmed for and computed by ACOS-6 computer, computer centre, Tohoku University.
The stress distribution $\tau(x, t)$ caused by dynamical piling-up is written as:

$$\tau(x,t) = \frac{A}{x-\ell} + A\sum_{i=2}^{n} \frac{1}{x-x_i} + \dot{\tau} t \quad , \tag{14}$$

which is shown in Fig.4.

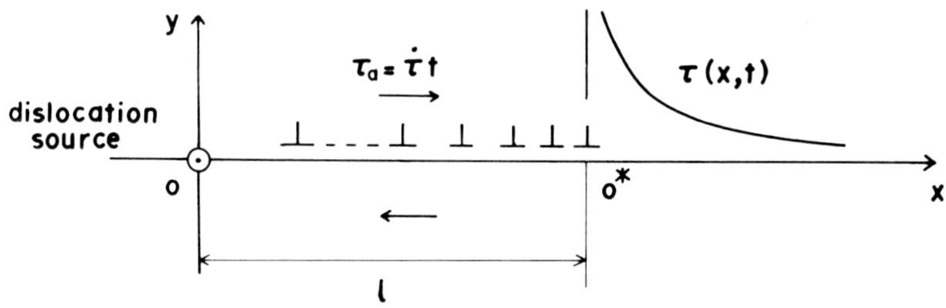

Fig.4. Stress distribution caused by dynamical piling-up of dislocations emitted from the source.

Then dynamic stress intensity factor by dislocation pile-up formation is written as:

$$K(t) \cong \sqrt{2\pi(x-\ell)}\ \tau(x,t)\Big|_{\ell < x < \ell(1+\varepsilon^*)} \ , \tag{15}$$

where $\ell\varepsilon^*$ is the small distance in which the stress distribution has $1/\sqrt{x^*}$ characteristic. $(x^* = x - \ell = \ell\varepsilon^*)$. If we use the non-dimensional variables, eqns. (14) and (15) reduce to, respectively:

$$\overline{\tau}(s_i, \theta) = \frac{A}{\dot{\tau}_0 t_0 \ell}\left(\frac{1}{s-1} + \sum_{i=2}^{n} \frac{1}{s-s_i}\right) + \frac{\dot{\tau}}{\dot{\tau}_0}\theta \quad , \qquad (s = x/\ell > 1)$$

$$\cdots\cdots (16)$$

and

$$\overline{K}(\theta) \cong \sqrt{s-1}\ \frac{\tau\ (x,t)}{\dot{\tau}_0 t_0} \tag{17}$$

For analysis of $\overline{K}(\theta)$, we take averaged values $\overline{K_m}(\theta)$of $\overline{K_q}(\theta)$ ($q=1\sim P$), over ε^* region with $1/\sqrt{s^*}$ singularity characteristic ($s^* = s - 1$) as follows:

$$\overline{K_m}(\theta) = \frac{1}{P}\sum_{q=1}^{P} \overline{K_q}\ (s,\theta) \tag{18}$$

On the other hand, the stress intensity factor usualy obtained on the concept of continuously distributed infinitesimal dislocations for the equilibrium piling-up is given as follows[19] for equal stress level:

$$K_s(t) = \sqrt{2\pi\ell}\ \tau_a \tag{19}$$

Thus it can be seen that $K_s(t)$ is static stress intensity factor.

Using the non-dimensional variables, eqn.(19) reduces to

$$\overline{K_s}(\theta) = \frac{K_s(t)}{\sqrt{2\pi\ell}\ \dot{\tau}_0 t_0} = \frac{\dot{\tau}}{\dot{\tau}_0}\theta \tag{20}$$

In calculation values for pureiron was used as material constant,[20] say, $m = 3$, $\tau_0{}^* = 6\ kg/mm^2$.

(2)-2 Results of calculations

The numerical results of stress distribution on the slip line ($s>1$) near the barrier is obtained by eqn.(16) as shown in Figs.5 and 6, where $s^*=s-1$, that is, the non-dimensional distance near the barrier. When the number of dislocations emitted is small as shown in Fig.5,the stress distribution near the barrier shows approximately $1/_{s^*}$ singularity closely near by the barrier and with increase of s^*,$1/\sqrt{s^*}$ singularity appears but is restricted within narrow region of $\alpha^*<s^*<\varepsilon^*$ ($\alpha^*\doteqdot 3\times10^{-3}$, $\varepsilon^*\doteqdot 2\times10^{-2}$). On the other hand, when the number of dislocations emitted increases much more as shown in Fig.6(sixty number of

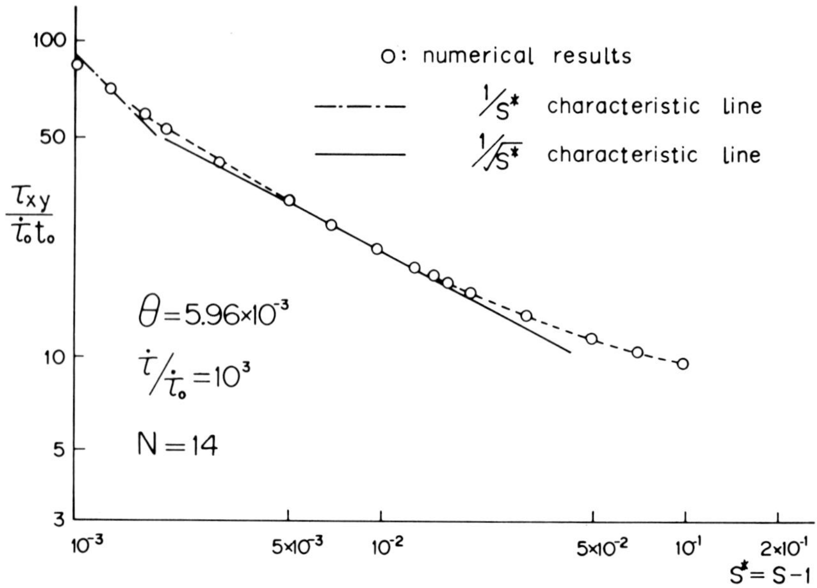

Fig.5. Stress distribution on the s* axis from the locked dislocation. For N=14.

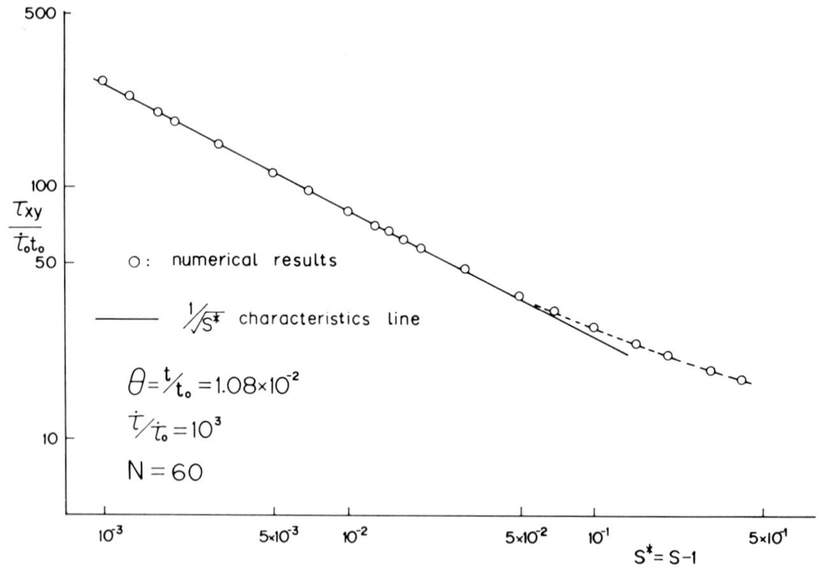

Fig.6. Stress distribution on the s* axis from the locked dislocations. For N=60.

dislocations emitted), the stress distribution shows $1/\sqrt{s^*}$ singularity from the vicinity of the barrier and this characteristic region extends up to five percent of slip line length. ($\varepsilon^* \doteqdot 5\times10^{-2}$, τ_a=9.85kg/mm^2, $\overline{K_m}(t)$=1.9kgmm$^{-3/2}$).

Numerical results of the dynamic stress intensity factor $\overline{K_m}(\theta)$ due to pile-up by dislocation group dynamics with emission is obtained as shown in Fig.7 by using eqns. (16),(17) and the $1/\sqrt{s^*}$ singularity of the stress distribution such as Figs.5 and 6 up to τ=9.85kg/mm^2 of applied stress level and sixty number of dislocations emitted. Static solution, $\overline{K_s}(\theta)$ obtained by the concept of continuously distributed infinitesimal dislocations for the equilibrium pile-up is also shown in Fig.7 for comparison with the dynamic stress intensity factor against equal stress level. $\overline{K_s}(\theta)$ and the linear part of $\overline{K_m}(\theta)$ in Fig.7 are written as follows,

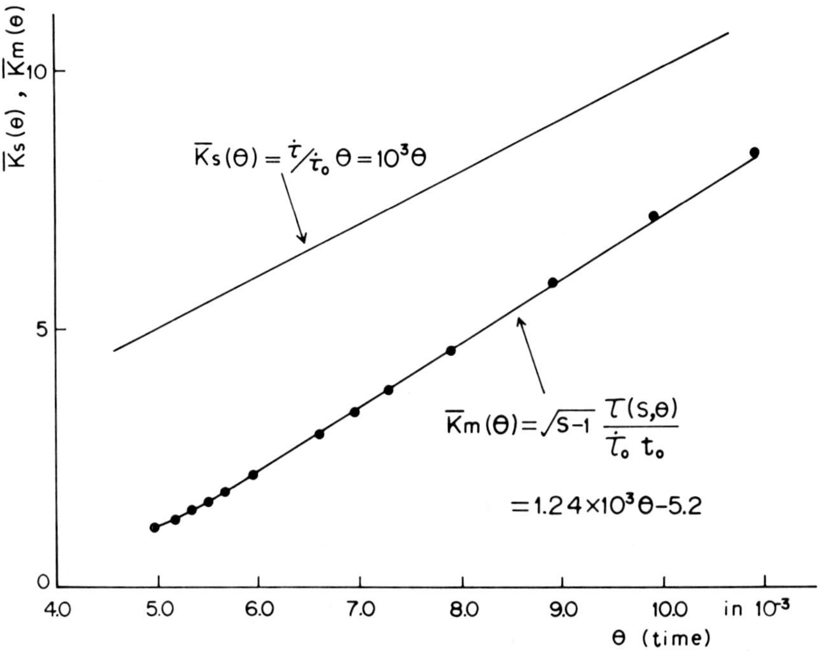

Fig.7. Dynamic stress intensity factor $\overline{K}m(\theta)$ as compared with equilibrium stress intensity factor $\overline{K}s(\theta)$.

$$\overline{K_s}\,(\theta) = 10^3\theta \tag{21}$$

$$\overline{K_m}\,(\theta) = 1.24\times10^3 - 5.2 \tag{22}$$

The ratio of $\overline{K_m}(\theta)$ to $\overline{K_s}(\theta)$ is plotted as a function of time θ as shown in Fig.8.

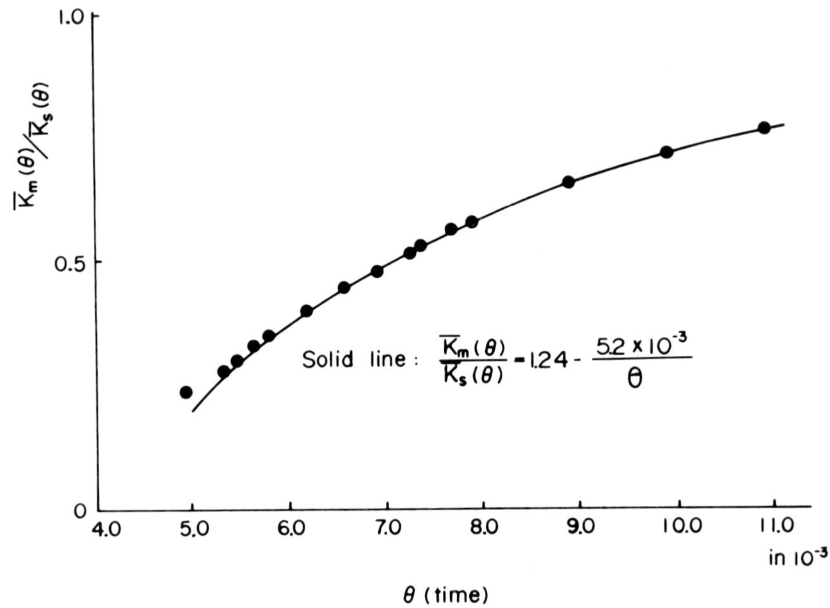

Fig.8. The ratio of $\bar{K}m(\theta)$ to $\bar{K}s(\theta)$ versus θ.

From these results it can be seen that the dynamic stress intensity factor is smaller than static one (equilibrium one) and appears to approach asymptoticaly to the static one as the number of dislocations emitted increases.

The ratio of the number of dislocations emitted under dynamic pile-up, N_P to that under free expansion motion, N_f[10)-13)] is calculated numerically as a function of Ⓗ $\left(\equiv\left(\frac{\dot{\tau}}{\dot{\tau}_0}\right)^{0.8}\theta\right)$ for pure iron[10)-13)] as shown in Fig.9. Ⓗ$_P$ corresponds to the value of Ⓗ at which the lead dislocation arrives at the barrier. It can be seen that N_P is much smaller than N_f due to back stress by pile-up dislocations exerted on the dislocation source.

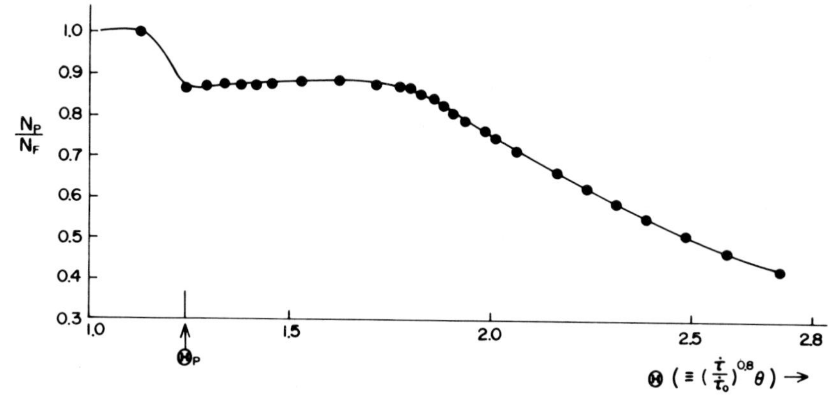

Fig.9. The ratio of the number of dislocations emitted until a specific time for dynamic pile-up, N_P to that for free emitting motion without a barrier, N_F plotted against dynamic factor Ⓗ .

(2)-3 Discussion

The ratio of the number of dislocations emitted until a specific time for the case of dynamic pile-up, N_p to that for the case of free emitting motion without a barrier decreases with increase of Ⓗ within large value of Ⓗ . By correlating the slip line length to grain size, it may be possible to apply this theory to the ductile fracture of polycrystalline.

From the results on $\overline{K_m}(\theta)$ it can be seen that the method proposed is possible to solve the problem for which the concept of continuously distributed infinitesimal dislocations cannot be applied because of small number of pile-up dislocations.

Furthermore, using analysis combining this with the global stress concentration near the crack tip, we can get microscopically local tensile stress so that the resultant local stress exceeds the atomic bonding force at the crack tip when combined with macroscopic stress concentration. In this way, the criterion for ductile fracture of this type may be obtained based on this model in terms of so-called combined micro and macro fracture mechanics.

Physico-phenomenological Model using Non-linearity in Crack Growth under Cyclic Stress at Not High Temperatures

Below or at room temperature, the plastic deformation near the fatigue crack is small under small scale yielding. Thus if we assume energy balance criterion only, the criterion determining crack growth rate may obey to linear elasticity fracture mechanics. On the other hand, fatigue crack growth rate is not predicted[3)-6)] completely by linear fracture mechanics. This, in turn, means that in fatigue crack propagation the local plastic stress distribution may play the dominant part. This may also be inferred since the plastic deformation near the crack tip may play important role whatever the physical model may be. Thus for the case of fatigue, even under small scale yielding it is necessary to formulate the local stress distribution near the crack tip. The result obtained is as follows:

$$\sigma_{\ell} = f(\beta)\,\sigma_{CY}\left(\frac{\Delta K}{\sqrt{\varepsilon}\,\sigma_{CY}}\right)^{\frac{2\beta}{1+\beta}}, \qquad (23)$$

where ΔK = stress intensity factor; σ_{CY} = the initial yield stress in cyclic straining (not static yield stress); β = the cyclic strain hardening exponent; $f(\beta)$ = some function of β; and ε = distant from the crack tip in the direction of crack.

By using eqn. (23) as local stress distribution near the crack tip, the mathematical formula for the fatigue crack growth can be derived for nucleation model[3)4)], for dislocation group dynamics model[5)], and for vacancy diffusion model[6)], respectively. $\frac{da}{dN}$ thus obtained is given by the form of the following form:

$$\frac{da}{dN} = C\left(\frac{\Delta K}{\sqrt{\varepsilon}\,\sigma_{CY}}\right)^{\mu}. \qquad (24)$$

μ is expressed by the equation as shown in Table 2 according to the different model. That is, $\frac{da}{dN}$ is in common with all three physical models as far as stress intensity factor dependence is concerned.

Table 2. Power coefficient μ of ΔK in da/dN, T = absolute temperature, k = Boltzmann's constant

Nucleation model	Dislocation group dynamics model	Vacancy diffusion model
$\frac{2\beta}{1+\beta}\left[\frac{1}{m'kT}+\frac{1}{2\beta}\right]$. $m'=\frac{1}{2.26\pi\rho}\left(\frac{v}{1-\nu^2}\right)^{-2/3}$ (Ref.3,4)	$\frac{2\beta}{1+\beta}\left[\frac{(m+1)^2}{(m+2)}+\frac{1}{2\beta}\right]$. $m=H_k/4kT$, H_k = activation energy (Ref.5)	$\frac{2\beta}{(1+p)(1+\beta)}\left[\frac{1}{m''kT}+\frac{1}{\beta}\right]$. m'' = constant, p = measure of time increase of the energy vacancy concentration in the plastic region (Ref.6)

As can be seen from Figs.5 and 6, as the number of pile-up dislocations emitted increase, the ε^* region in eqn(16) which covers $1/\sqrt{s^*}$ singularity extends from the vicinity of the barrier over to the five percent length of slip line. This length, ε^* may correspond to ε in eqns.(23) and (24), over which high local stress is averaged.

Conclusions

The physico-phenomenological models presented in this article may be in fairly general as they are phenomenological. In turn, the next step is to improve them more physical and more practicaly convenient and to correlate the models with more detailed or microscopic mechanisms in order to examine which is more crrect or to develop a new model.

Acknowledgements

The authors wish to thanks Dr. A. Kamei for his useful discussions. They also are grateful to Mr.H. Nishi for his help in calculation.

† On the other hand, atomic cohesive force in usual materials is far much larger than yield stress, σ_Y, for instance, 100 times.

References

1 Yokobori, T.: Criteria for Nearly Brittle Fracture. Int. J. Fract. Mech. 4(1968) 179.

2 Yokobori, T.; Maekawa, I.; Yokobori, A.T., Jr.; Sakata, H.: Constitutive Equations and Global Criteria for Ductile Fracture. Proc. IUTAM Symposium on Three-dimensional Constitutive Relationships and Ductile Fracture, June 2-5 (1980) Dourdan. (To be published)

3 Yokobori, T.: A Kinetic Approach to Fatigue Crack Propagation, Physics of Strength and Plasticity, A.S. Argon, ed., MIT Press (1969) 327-338.

4 Yokobori, T.; Ichikawa, M.: Effect of Elastic Plastic Stress, Distribution near the Crack on the Nucleation Theory of Fatigue Crack Propagation. Rep. Res. Inst. Strength and Fracture of Materials, Tohoku Univ. Sendai, Japan 4, (1968) 45-53.

5 Yokobori, T.; Konosu, S.; Yokobori, A.T., Jr.: Micro and Macro Fracture Mechanics Approach to Brittle Fracture and Fatigue Crack Growth. Proc. 4th Int. Conf. Fracture 1 (1977) 665-682.

6 Yokobori, T.; Ichikawa, M.: A Kinetic Theory of Fatigue Crack Propagation Based on Vacancy Condensation Mechanism. Rep. Res. Inst. Strength and Fracture of Materials, Tohoku Univ. Sendai, Japan 6 (1970) 75-86.

7 Yokobori, T.: Zairyo Kyodogaku, 2nd edition, (in Japanese), Iwanami Shoten (1974) 139; Russian edition, Naukova Dumka, Kiev (1978) 135.

8 Ludwik, P.; Scheu, R.: Ueber Kerbwirkungen bei Flußeisen. Stahl und Eisen 43 (1923) 999-1001.

9 Cottrell, A.H.: Fracture Proc. Roy. Soc. London A 276 (1963) 1-18.

10 Yokobori, T.; Yokobori, A.T., Jr.; Kamei, A.: Computer Simulation of Dislocation from a Stressed Source. Phil. Mag. 30 (1974) 367-378.

11 Yokobori, A.T., Jr.; Yokobori, T.; Kamei, A.: Generalization of Computer Simulation of Dislocation Emission under Constant Rate of Stress Application. J. Appl. Phys. 46 (1975) 3720-3724.

12 Yokobori, A.T., Jr.; Kawasaki, T.; Yokobori, T.: Computer Simulation of Dislocation Groups Dynamics Under Applied Constant Stress and Its Application to

Yield Problems of Mild Steel. High Velocity Deformation of Solids, edited by Kawata, K. and Shioiri, J., IUTAM Symposium Tokyo/Japan, 1977 (1979) 132-148.

13 Yokobori, A.T., Jr.; Yokobori, T.; Kamei, A.: Computer Simulation of the Dynamic Behaviour of Dislocation Groups of Opposite Sign Emitted from a Stressed Source. J. Materials Science and Engineering 40 (1979) 111-118.

14 Yokobori, T.: Zairyo Kyodogaku, 2nd edition, (in Japanese), Iwanami Shoten (1974)112.

15 Yokobori, T.; Konosu, S.: The Elastic-Plastic Stress Distribution near the Flat Surfaced Notch with Cycloidal Tip in a Work-hardening Material. Ing.-Arch. 45 (1976) 243-247.

16 Cottrell, A.H.: Dislocations and Plastic Flow in Crystals. Clarendon Press, Oxford (1953).

17 Kanninen, M.F.; Rosenfield, A.R.: Dynamics of Dislocation Pile-up Formation. Phil. Mag. 20 (1969) 569-587.

18 Gerstle, F.P.; Dvorak, G.J.: Dynamic Formation and Release of a Dislocation Pile-up against a viscous Obstacle. Phil. Mag. 29 (1974) 1337-1346.

19 Yokobori, T.; Ichikawa, M.: The Interaction of Parallel Elastic Cracks and Parallel Slip Bands Respectively Based on the Concept of Continuous Distribution of Dislocations I. Rep. Res. Inst. Strength and Fracture of Materials, Tohoku Univ., Japan 3 (1967) 1-14.

20 Turner, A.P.L.; Vreeland, T.,Jr.: The Effect of Stress and Temperature on the Velocity of Dislocations in Pure Iron Monocrystals. Acta Metallurgica 18 (1970) 1225-1235.